THE GOLDILOCKS POLICY

The Basis for a Grand Energy Bargain

THE GOLDILOCKS POLICY

The Basis for a Grand Energy Bargain

John R Fanchi

Texas Christian University, USA

World Scientific

NEW JERSEY • LONDON • SINGAPORE • BEIJING • SHANGHAI • HONG KONG • TAIPEI • CHENNAI • TOKYO

Published by

World Scientific Publishing Co. Pte. Ltd.
5 Toh Tuck Link, Singapore 596224
USA office: 27 Warren Street, Suite 401-402, Hackensack, NJ 07601
UK office: 57 Shelton Street, Covent Garden, London WC2H 9HE

British Library Cataloguing-in-Publication Data
A catalogue record for this book is available from the British Library.

THE GOLDILOCKS POLICY
The Basis for a Grand Energy Bargain

Copyright © 2019 by World Scientific Publishing Co. Pte. Ltd.

All rights reserved. This book, or parts thereof, may not be reproduced in any form or by any means, electronic or mechanical, including photocopying, recording or any information storage and retrieval system now known or to be invented, without written permission from the publisher.

For photocopying of material in this volume, please pay a copying fee through the Copyright Clearance Center, Inc., 222 Rosewood Drive, Danvers, MA 01923, USA. In this case permission to photocopy is not required from the publisher.

ISBN 978-981-3276-39-0
ISBN 978-981-3277-44-1 (pbk)

For any available supplementary material, please visit
https://www.worldscientific.com/worldscibooks/10.1142/11159#t=suppl

Printed in Singapore

To my grandsons Cameron and Caleb, and future generations

Preface

This book presents a roadmap to sustainable energy. Society is in a period of transition from one energy mix to another. We are witnessing a passionate debate between advocates of competing energy strategies on many levels: technological, political, and geopolitical. In this book, we make the case for a grand energy bargain that recognizes the need to protect the environment from the combustion of fossil fuels while protecting national and global economies during the transition from fossil fuels to sustainable energy.

The Goldilocks Policy: The Basis for a Grand Energy Bargain is written for anyone who consumes energy. You will learn

- what energy is and why energy is important to our quality of life
- that we should consider changing our energy mix because our current energy mix relies on finite resources that may be affecting our environment
- that we have historical guidance for establishing a period of transition to a new energy mix
- about a plan called the Goldilocks Policy for energy transition that is designed to guide the transition to a sustainable energy portfolio that provides cost-effective energy while protecting the environment

- that several domestic and global obstacles must be overcome before a comprehensive energy policy can succeed
- what the future of energy could be if we have the discipline and patience to implement the Goldilocks Policy.

I thank K. Wayne Fanchi for useful comments.

John R. Fanchi, Ph.D.
November 2018

Contents

Preface		vii
Part I	**Introduction**	
	1. What is energy?	2
	2. How much energy do we consume?	3
	3. Why should we care about energy?	11
	4. Is our use of energy affecting the climate?	14
	5. Should we be concerned about our supply of energy?	17
Part II	**Is the Climate Change Debate Settled?**	
	6. Is our climate changing?	24
	7. What is affecting the atmosphere?	25
	8. Are we changing the atmosphere?	29
	9. Has climate change occurred before? The long view	41
	10. What might be the consequences of anthropogenic climate change?	46
	11. How can we reduce greenhouse gas emissions?	48
	12. Is the climate change debate settled?	52
	12.1 Population growth	54
	12.2 IPCC versus NIPCC: Is this the debate?	55
	12.3 COP21: The Paris climate meeting	58
	12.4 The oil and gas climate initiative	60

Part III Roadmap to a Sustainable Energy Mix

13. Trend toward decarbonization 64
14. Competing energy visions 65
15. How can we transition to a sustainable energy mix? 69
    15.1 John Hofmeister and regulation 69
    15.2 Vaclav Smil and the transition from fossil fuels to renewables 70
    15.3 Daniel Yergin and the quest for change 73
16. Goldilocks policy for energy transition 75
    16.1 Historical basis for duration of energy transitions 75
    16.2 Temperature change forecast 77
    16.3 The Goldilocks policy 79

Part IV Obstacles to Adopting the Goldilocks Policy

17. Conventional political obstacles 84
    17.1 Typical forms of government 84
    17.2 Government and energy 86
    17.3 Mass media 86
18. Role of geopolitics 87
    18.1 Clash of civilizations 87
    18.2 Clash over resources 93
    18.3 Energy interdependence 93
19. The political roots of socialist environmentalism 95
    19.1 Is modern environmentalism an attempt to impose socialism? 95
    19.2 Dialectical materialism 97
    19.3 The Marxist view of property 99
    19.4 The Marxist view of natural resources 100
    19.5 The Marxist view of environmentalism goes national 102

		19.6 The Fabian Society: Evolutionary rather than revolutionary	104
		19.7 The Marxist view of environmentalism goes global	105
	20.	Oligarchic political obstacles	106
		20.1 Quigley's plutocracy model	107
		20.2 The bureaucratic ruling class model	115
	21.	Fabian socialism as a political obstacle	120
		21.1 The Fabian Society	120
		21.2 Fabian globalism	124
	22.	Globalism as a political obstacle	126
		22.1 World War I and The Inquiry	126
		22.2 Walter Lippmann's brush with Fabian socialism	128
		22.3 Wilson's 14 Points and the League of Nations	130
	23.	Environmentalism and the United Nations	132
	24.	Maurice Strong and global socialist environmentalism	135
	25.	International banking and globalization	147
	26.	Funding of globalization by the privileged minority	151
		26.1 Amassing the Rockefeller fortune	151
		26.2 The Rockefeller Foundation	153
		26.3 International banker David Rockefeller	157
	27.	Picking up the environmental baton: Barack Obama	166
Part V	**What Is the Future of Energy?**		
	28.	Can the obstacles be overcome?	174
		28.1 Competing economic visions	175
		28.2 Global feudalism	177
	29.	Selecting our energy future	178

Appendix A: The Goldilocks Story 181
Appendix B: The Earth Charter 183
References 195
Index 209

Part I
Introduction

1. What is energy?

Energy is the primary focus of our discussion, so it is worthwhile spending a little time clarifying what we mean by energy. Energy is the ability to do work. There are many forms of energy. Two of the most common forms of energy are potential energy and kinetic energy. Potential energy is the ability to produce motion, and kinetic energy is the energy of motion. Other forms of energy include heat (thermal energy), light (radiant energy), chemical energy, electrical energy, biological energy, nuclear energy, and gravitational energy.

An important characteristic of energy is its ability to transform from one form of energy to another. For example, mechanical energy of motion can be transformed into electrical energy using a generator, while nuclear fission plants convert mass to electrical energy.

Fossil energy, nuclear energy, and renewable energy are considered primary energy sources because they can provide energy directly. Carriers of energy are produced from primary sources of energy and include electricity and hydrogen. Both electricity and hydrogen are considered "carriers" of energy because they can be used to store and deliver energy in a useful form.

There are many sources of energy, but they are not all capable of providing energy on a scale that is needed either nationally or globally. The viability of an energy source as a large scale source of energy needs to be considered in any analysis of its suitability as a key component in an energy mix. The viability of an energy source is not just a question of quantity available, but is also a question of how that energy can be captured and delivered to an end user efficiently and cost-effectively. An extensive discussion of different energy issues, including the relative merits of different energy types, is provided in Fanchi and Fanchi [2017]. We review the importance of energy and the need to undertake a transition to a sustainable energy mix in the rest of Part I.

Climate change is discussed in Part II. Although the question of human involvement in climate change is still unsettled, enough data is available to motivate a transition away from fossil fuels. The transition does not have to be abrupt and catastrophic, however. Historical energy transitions can be a guide to a reasonable duration for making an orderly transition.

Our future energy mix depends on choices we make, which depends, in turn, on energy policy. Competing visions for reaching a sustainable energy mix and the Goldilocks Policy for Energy Transition are discussed in Part III. If the energy transition is too fast, it could significantly damage the global economy. If the energy transition is too slow, damage to the environment could be irreversible. The Goldilocks Policy for Energy Transition is designed to establish a middle ground between competing visions. The Goldilocks Policy is named after the Goldilocks story, which is presented in Appendix A. The essence of the Goldilocks story is that the best option is often bounded by less desirable options.

We show in Part III that the duration of the energy transition should be just right, and we describe how to estimate the amount of time we have to make a transition to a sustainable energy mix. We need to adopt a reasonable plan of action that reduces uncertainty for businesses and innovators with predictable public policy while simultaneously minimizing environmental impact.

If we exercise strategic vision, discipline, and patience, we can overcome the obstacles to successful implementation of a grand energy bargain. There are many obstacles that stand in the way of successful implementation of the Goldilocks Policy for Energy Transition. Several key obstacles are discussed in Part IV. We summarize our observations and conclusions in Part V.

2. How much energy do we consume?

World energy consumption from 1980 to 2015 is shown in Figure 2-1. The figure shows the global reliance on fossil fuels, which includes coal, oil and natural gas. The world consumed approximately 575 Quads of energy in 2015, and the United States consumed approximately 98 Quads in 2015 [US IEO, 2018]. A Quad is one quadrillion BTU of energy. One BTU (British Thermal Unit) of energy is the amount of heat energy needed to raise the temperature of one pound of water 1°F. A BTU is equivalent to 1055 J of energy in the metric system. The Quad is a unit of energy that is suitable for expressing energy consumption and production on a global scale.

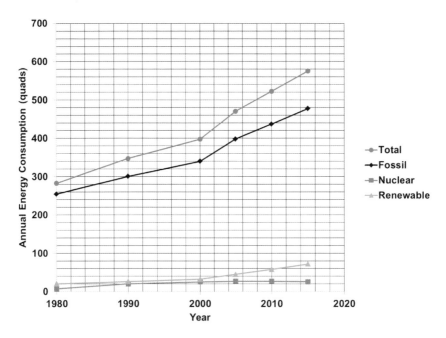

Figure 2-1. World Annual Energy Consumption by Fuel Type from 1980 to 2015. [US IEO, 2018]

We can categorize energy sources as either nonrenewable or renewable. Nonrenewable energy is energy that is obtained from a source at a rate that exceeds the rate at which the source is replenished. By contrast, renewable energy is energy that is obtained from a source at a rate that is less than or equal to the rate at which the source is replenished. Examples of nonrenewable energy sources include fossil fuels and nuclear fuel such as uranium, while renewable energy sources include hydropower, wind, and solar energy.

Table 2-1 shows that the global reliance on fossil fuels has continued into the 21st century. Nuclear electric power relies on nuclear fission technology, which is the decay of large radioactive nuclei such as uranium into smaller nuclei. The reliance on nuclear fission power is important because it differs substantially from the other source of nuclear energy: nuclear fusion. Nuclear fusion energy results from the release of energy when smaller nuclei merge to form a larger nucleus. Nuclear fusion is the process that powers the Sun, and may be the solution to all of our energy problems. Nuclear fusion energy is not yet commercially viable, but it is expected to become an energy source in this century. We discuss nuclear fusion in more detail later.

Table 2-1. Annual World Energy Consumption by Source, 1980–2015. [US IEO, 2018]

Energy Source	Energy Consumption (as % of total)					
	1980	1990	2000	2005	2010	2015
Total	100.0	100.0	100.0	100.0	100.0	100.0
Fossil Fuels	90.0	86.6	85.5	84.5	83.6	83.0
Nuclear Electric Power	2.7	5.9	6.4	5.8	5.2	4.5
Renewable Energy	7.3	7.6	8.1	9.6	11.1	12.5
Total (quads)	283.2	347.6	397.9	471.0	522.9	575.4

The history of energy consumption in the United States from 1775 to 2015 is shown in Figure 2-2. It illustrates energy consumption in most of the developed world over that period of time. Wood, a combustible biomass, was the primary energy source until the discovery of coal as a combustible material in the mid-1800's. The energy category called "Other" includes conventional hydroelectric power, geothermal, solar thermal, photovoltaic, and wind. Petroleum, once known as "rock oil" to distinguish it from "whale oil," became an important contributor to the U.S. energy mix in the second half of the 19th century. Nuclear fission energy became a commercial energy source following the end of World War II in 1945. It took a few years to develop peaceful electricity generating technology based on nuclear fission reactions. Hydroelectric energy and wood are the only renewable energy sources that made a major contribution to the U.S. energy mix prior to the 21st century.

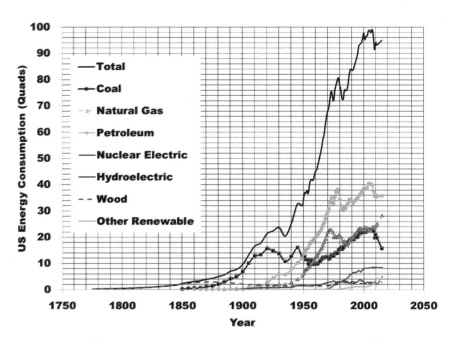

Figure 2-2.   U.S. Energy Consumption by Source, 1775–2015 (Quadrillion BTU). [US EIA, 2018]

Figure 2-3 shows per cent contribution of different energy sources to the U.S. energy mix. Each peak shows when a new energy source began to take over market share from a previous energy source. Coal, for example, began to replace wood around 1850 and peaked in the early 1900s. Oil, natural gas, and hydroelectric energy began to take over market share from coal by 1900.

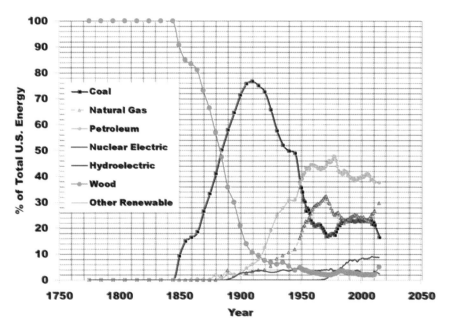

Figure 2-3. U.S. Energy Consumption by Source, 1775–2015 (% of Total).

Historically, the replacement of one energy source by another had a beneficial environmental impact at the time of the replacement. For example, the discovery of coal as a combustible material helped prevent the deforestation of England during the industrial revolution. Demand for nighttime illumination by Americans in the middle of the 19th century exceeded the four million barrels of whale oil produced each year by refineries in New England and New York. It was for this reason that "Colonel" Edwin Drake's Connecticut backers sent him to Oil Creek Valley, Pennsylvania to drill for oil. The importance of Drake's well is shown in an 1861 cartoon from *Vanity Fair* (Figure 2-4). The cartoon shows whales celebrating the discovery that "rock oil" could replace "whale oil" as a combustible fuel. It is fair to say that, at the time, oil saved the whales.

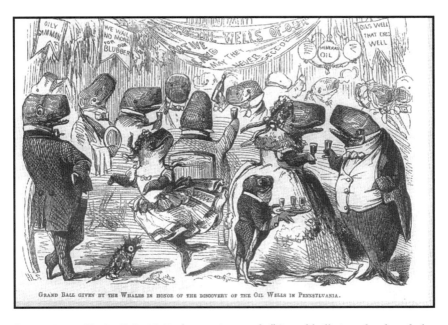

Figure 2-4. *Vanity Fair*, 1861: the caption reads "Grand ball given by the whales in honor of the discovery of the oil wells in Pennsylvania."

Figure 2-5 shows more detail about the contribution of different energy sources to the U.S. energy mix in 2015. Crude oil was the largest component, and geothermal was the smallest component. Biomass (organic material such as wood) and hydropower dominate the renewable energy sector, followed by wind and geothermal. As a percentage of the overall energy pie, renewables are still a relatively minor contributor.

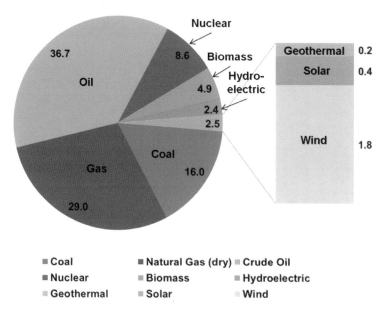

Figure 2-5. U.S. Energy Consumption by % Energy Source at End of 2015. Total Annual U.S. Energy Consumption = 97.1 Quadrillion BTU. [U.S. EIA, 2018]

Figure 2-6 shows how energy was distributed in 2015 between the four demand sectors of the U.S. economy. The term "demand sector" is used to define the broad segments of society which consume energy resources: transportation, industrial, residential and commercial, and electric power. The first number in each box is the amount of energy in Quads and the parenthetic number is percent of total. The percent of source and percent of sector values are shown adjacent to the lines between the boxes. For example, 100% of nuclear fission energy was used to produce 22% of electric power in 2015. The rest of the electric power was provided using coal, natural gas, renewables, and oil. Petroleum (liquid hydrocarbons such as gasoline) provided most of the energy to the transportation sector in 2015. By contrast, coal provided most of the energy to utilities which generate electric power in 2015.

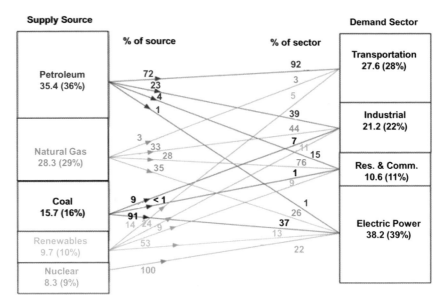

Figure 2-6. U.S. Primary Energy Consumption by Source and Sector, 2015 (%). [US EIA, 2016]

A study of energy flow from energy sources to end users shows that over half of the energy supply is lost energy, much of it in the form of heat. Attempts to improve the efficiency of energy use can substantially reduce the amount of energy that needs to be supplied to the economy. Energy conservation is designed to increase energy use efficiency.

An energy mix can be categorized as sustainable or unsustainable. A sustainable energy mix is the mix of energy resources that allows society to meet its present energy needs while preserving the ability of future generations to meet their needs. This concept is based on the sustainable development concept developed by the 1987 UN World Commission on Environment and Development. The commission, which was led by Ms. Gro Harlem Brundtland of Norway, adopted a principle of sustainable development that asserted that the present generation should meet its present needs while preserving the ability of future generations to meet their needs. An unsustainable energy mix is an energy mix that does not satisfy the sustainability requirement.

3. Why should we care about energy?

Traditional producers of energy are no longer part of an Oil and Gas Industry, or Coal Industry, but have become part of a highly competitive Energy Industry. These changes are affecting what we pay for energy as producers of competing energy sources and their allies vie for dominance in the emerging energy mix.

Long-term forecasts of energy production and consumption assume that the energy mix will change substantially during a human lifetime. This raises questions such as the following: How will the changing energy mix impact our lives? How will it impact national and global economies? How will it impact the environment? The answers to questions like these are partly within our influence, and will certainly affect our lives.

We can show the importance of energy to our quality of life by comparing the United Nations Human Development Index (UN HDI) to energy consumption. Three factors are used to determine the UN HDI for each nation: life expectancy at birth; education measured by mean years of schooling and the expected years of schooling; and the standard of living represented by Gross National Income per capita.

The UN HDI is a measure of the quality of life on a national level. It is a number that ranges from 0 (minimal human development) to 1 (maximum human development). Values of the UN HDI that are close to one are associated with countries with a relatively high quality of life, while countries with a UN HDI value closer to zero have a relatively low quality of life. Figure 3-1 illustrates the global distribution of UN HDI. Darker colors in Figure 3-1 are higher HDI values, while lighter colors are lower HDI values. As a rule, countries in the developed world have a higher UN HDI than countries in the less developed world. Furthermore, UN HDI is correlated to energy consumption.

12 | The Goldilocks Policy

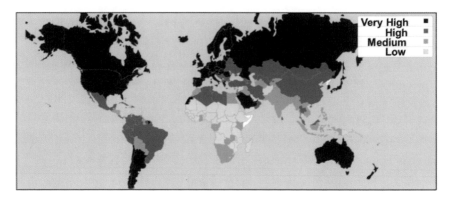

Figure 3-1.   UN HDI Map (2016 data). [UNDP 2016 HDR, 2017]

A plot of UN HDI as a function of per capita primary energy consumption is shown in Figure 3-2 for 2015 data [UN 2015 HDI, 2018; US 2015 EIA, 2018], which was when all of the data was available.

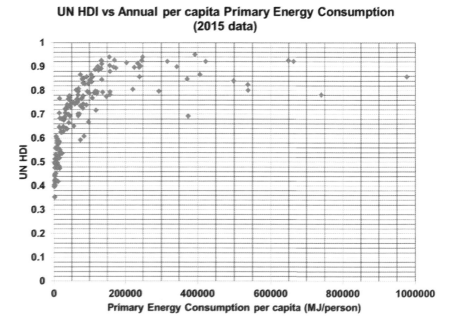

Figure 3-2.   United Nations HDI versus Annual Per Capita Primary Energy Consumption (2015 Data). [UN 2015 HDI, 2018; US 2015 EIA, 2018]

Per capita energy consumption is expressed in megajoules per person (MJ per person), which is a measure of energy use. The UN HDI tends to increase as per capita electricity consumption increases. It begins to level off when per capita electricity consumption exceeds approximately 200,000 MJ per person.

Smil performed an analysis of the correlation of the quality of life components of the UN HDI with energy use [Smil, 2003, pages 97–105]. Smil used the definition of UN HDI available at the time, which included Gross Domestic Product per capita rather than the more modern Gross National Income per capita. Smil pointed out that correlations between measures of quality of life and energy consumption are not perfect, but they are indicative of the amount of energy needed to achieve a relatively high UN HDI above 0.8 [Smil, 2003, page 102]. He concluded "that the average consumption of between 50–70 GJ/capita (equivalently, 50,000–70,000 MJ/capita) provides enough commercial energy to secure general satisfaction of essential physical needs in combination with fairly widespread opportunities for intellectual advancement and with respect for individual freedoms." [Smil, 2003, page 352]. He observed that energy consumption to achieve good health, as measured by indicators such as infant mortality and life expectancy at birth, requires approximately 110,000 MJ/capita [Smil, 2003, page 351].

Smil expressed concern in 2003 that "shaping the future energy use in the affluent world is primarily a moral issue, not a technical or economic matter. So is the narrowing of the intolerable quality-of-life gap between the rich and the poor world." [Smil, 2003, page 370] Over a decade later, after reviewing the history of energy and civilization, Smil pointed out that civilizations have been characterized by expansion and increasing complexity. He asked "Can we reverse these trends by adopting the technically feasible and environmentally desirable shift to moderated energy uses?" [Smil, 2015, page 440] Countries with relatively high UN HDI values, such as Iceland and the USA, may be able to reduce their per capita electricity consumption by increasing energy use efficiency while still preserving their quality of life. If sufficient moderation of energy use does not occur, a review of the relationship between energy consumption per capita and quality of life measures such as the UN HDI shows that we should expect global energy demand to increase as countries with relatively low UN HDI values strive to increase their quality of life.

4. Is our use of energy affecting the climate?

Many people believe that our current reliance on fossil fuels is having an adverse environmental impact. They believe that emissions from the combustion of fossil fuels are contributing to climate change. The theory of climate change assumes that the combustion of carbon-based fuels increases the concentration of greenhouse gases in the atmosphere. Greenhouse gases like water vapor and carbon dioxide absorb infrared radiation and warm the atmosphere.

The combustion of fossil fuels produces some byproducts that pollute. For example, the chemical combustion of hydrocarbon fuel and oxygen produces heat, carbon dioxide, and water. Carbon dioxide and water vapor are greenhouse gases because they absorb infrared radiation, or heat energy, that passes through the atmosphere. Other greenhouse gases include methane, nitrous oxide, and volatile organic compounds.

Carbon dioxide is emitted into the atmosphere by combustion and by exhalation. Measurements of atmospheric carbon dioxide at Mauna Loa, Hawaii by Charles David Keeling and colleagues show that the concentration of carbon dioxide in the atmosphere is increasing. When Keeling began his measurements in the late 1950s, the carbon dioxide concentration was between 310 ppm and 320 ppm. On May 11, 2013 the concentration of carbon dioxide at Mauna Loa, Hawaii was measured to be 400 ppm. The Keeling curve is used by proponents of climate change as evidence that our reliance on carbon-based fuels is continuing to increase the concentration of greenhouse gases in the atmosphere.

The Keeling method has been used at locations around the world to see if carbon dioxide concentrations are following the same pattern as the Mauna Loa Keeling curve. The United States National Oceanographic and Atmospheric Administration (NOAA) has made the data publicly available [US NOAA Keeling, 2015]. The data cover a shorter period of time than the Mauna Loa data, but plots of the data at locations around the world show annual variations of CO_2 concentration and a steadily increasing average similar to the pattern exhibited by the Mauna Loa Keeling curve. Figure 4-1 presents an overlay of CO_2 concentration data from relatively isolated monitoring stations at Mauna Loa, Hawaii; Barrow, Alaska; Summit, Greenland; Tutuila, American Samoa; and South Pole, Antarctica.

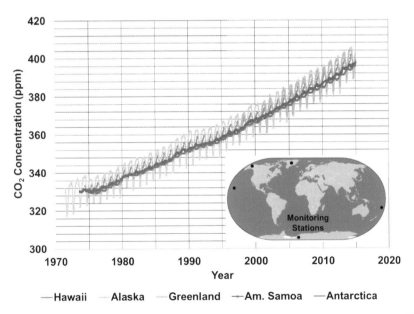

Figure 4-1. Overlay of CO_2 Concentration from Different Parts of the World. [US NOAA Keeling, 2015]

The Keeling curve reinforces other measurements of atmospheric carbon dioxide that indicate a rapid increase in atmospheric greenhouse gas concentrations during the past two centuries. This increase is concurrent with increases in carbon-based fuel consumption, beginning with coal and the Industrial Revolution. It is also concurrent with a significant growth in global population.

Before the term "climate change" was adopted to describe adverse changes to the climate caused by human activity, the term "global warming" was used. Before global warming, some scientists expressed concern about global cooling as we entered an Ice Age. During the Cold War, there was also a concern that a nuclear war would lead to "nuclear winter." Nuclear winter refers to a decline of air temperature as a result of an increase in atmospheric particulates following the detonation of many nuclear weapons.

A coterie of modern scientists argue that an increase in atmospheric greenhouse gases can lead to atmospheric warming. Measurements of atmospheric temperature change have shown that atmospheric temperature is no longer increasing, and has stabilized in recent years. Global warming was no longer an accurate term, and the term climate change was introduced at the beginning of the 21st century. Today, extreme weather events such as hurricanes, droughts, and tornadoes are considered evidence of climate change. If the climate is in fact changing, however, the question remains: what proportion of this change is due to human activity?

Some scientists attribute virtually all climate changes to human activity, while other scientists argue that the impact of human activity has been overstated. For example, meteorologist Brian Sussman presented a summary of the global temperature record commencing with the most recent Ice Age and proceeding to the present [Sussman, 2010, pages 60–61]. This period covered the Medieval Warm Period from 900–1300 A.D., the Little Ice Age from 1350–1800 A.D., and temperature changes from 1850 to the present. Sussman argued that a compilation of temperature records indicated a warming of approximately 1°F since the mid-nineteenth century. He suggested that there has been a net cooling since 1930. With reputable voices on both sides of the debate, we are left to wonder: Has the debate about climate science been settled? We consider this question in more detail in Part II.

5. Should we be concerned about our supply of energy?

Global oil production rate has been between 70 and 80 million bopd (barrels of oil per day) for most of the 21st century. This rate has been enough to meet demand. Quality of life depends on energy consumption, and the desire to improve quality of life is increasing demand for energy. The demand for energy early in the 21st century raised concerns that the supply of fossil fuels, especially oil and gas, is not sufficient to meet demand indefinitely. A concern about the supply of oil is based in part on the widespread belief that fossil fuels are finite resources.

Oil and gas are carbon-based fuels that are made primarily from carbon and hydrogen, with small quantities of other chemicals such as sulfur, nitrogen, oxygen, and metals such as iron, calcium, and potassium. The elemental composition of oil and gas is similar to the elemental composition of biological organisms — both depend on carbon and hydrogen atoms — which provides evidence about the origin of oil and gas. Most scientists believe that oil and gas are formed by the death and decay of biological organisms.

The leading theory of oil and gas formation is the biogenic theory, which asserts that the death, burial and subsequent decay of biological organisms leads to the formation of oil and gas. For example, plankton and other small organisms die at an oceanic reef and sink to the seabed where they can be buried. The temperature and pressure of a formation increases as depth increases. Combining heat and pressure with bacterial action transforms the organic matter into hydrocarbon molecules of varying sizes and weights. The mixture of hydrocarbon molecules can exist in the solid, liquid, or gas phase and can be trapped underground in a geological formation.

Based on the assumption that the biogenic theory is correct, scientists speculate that the process of forming hydrocarbons is very slow. This suggests that oil and gas being extracted and used for energy is being used at a rate that is much faster than the rate at which it can be replenished. Consequently, oil and gas are considered finite, nonrenewable resources.

The production of a finite natural resource will reach a point in time when half of the resource has been recovered. M. King Hubbert developed a technique in 1956 to predict the rate of oil production in the 48 contiguous states of the United States as a function of time. Hubbert found that the accumulation of oil production rates for all of the wells in a region would result in a Gaussian curve. Production is symmetric about the peak of a Gaussian curve. Peak oil supply is the point in time when maximum oil production rate is attained and half of the oil in the region is produced.

Hubbert applied his technique to production from the 48 contiguous states of the United States in 1956. He predicted a peak oil production rate of 2.7 to 3.0 billion barrels of oil per year during the peak year of 1970. A comparison of actual production to Hubbert's prediction showed that Hubbert's technique worked for conventional oil production in the 48 contiguous states.

Hubbert made a similar prediction of global oil production that same year. Hubbert's technique predicted that global oil production would peak at an oil production rate of 12 to 13 billion barrels of oil per year during the peak year of 2000.

Figure 5-1 shows that global oil production has not yet reached its peak. The Gaussian curves indicate that global oil production should peak around 2030. Some proponents of Hubbert's technique argue that global peak oil production has already been reached. They point out that the number of large fields being discovered is decreasing, and that Hubbert's prediction was not quite correct because of technological advances that made it possible to produce more oil than was extracted using older technology. Fossil fuel production has become more complicated technologically as oil and gas fields are being discovered in more remote locations, harsh environments, and complex structures.

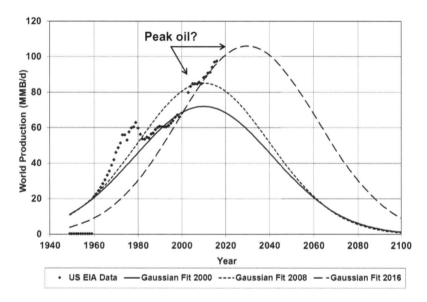

Figure 5-1. Forecast for Oil Produced Through 2016 Using Gaussian Curve. [US EIA, 2018]

Technological advances have enlarged global commercial oil and gas resources. The advances include the ability to drill in different directions rather than just vertically downward, to fracture very low permeability formations so fluids can flow to wellbores, and to obtain better images of the subsurface so that bypassed resources and new fields can be discovered. These advances make it possible to produce oil remaining in fields, and to produce more difficult resources in low permeability formations, notably oil and gas production from shale. The increase in oil production during the first two decades of the 21st century is largely due to shale oil production in the United States.

New technology is making it possible to produce more oil from conventional fields, and oil that would have been considered uneconomic in 1956 when Hubbert published his work. Despite advances in technology, the consensus opinion is that carbon-based fossil fuels are finite, and that they will not be available on an indefinitely sustainable basis.

If we look at global oil production on a per capita basis, the peak in global oil production was reached in the 1970s. Figure 5-2 shows that the increase in global population is making it more difficult to maintain oil production rate per capita. It does not appear that fossil energy will be able to meet increasing energy demand as the population grows and nations seek to attain a higher quality of life for their people.

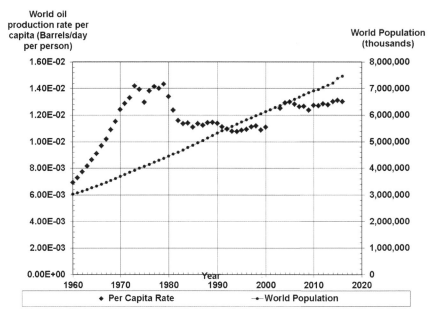

Figure 5-2. Did Oil Production Rate per Capita Peak Before 1980?

Alternative sources of fuel need to be developed. Substitutes for fossil energy include nuclear energy, wind energy, solar energy, energy from tides and waves, and bioenergy. Most of these substitutes are more expensive than fossil energy, but have other benefits that their proponents believe are sufficient to justify the additional expense. For example, a wind turbine can generate electrical energy without producing greenhouse gases. Wind is not totally free of greenhouse gas production; the acquisition of resources for the manufacture of wind turbines, in addition to the manufacturing process itself, presently require the use of carbon-based fuels. Greenhouse gases and their impact on the environment are discussed in more detail later.

The relative cost of electricity generation is shown in Figure 5-3. The figure is based on US EIA estimates [US EIA AEO, 2017] of the Levelized Cost of Electricity (LCOE) and compares estimates using 2009$ [US AEO, 2011] and 2016$ [US AEO, 2017]. The LCOE is the sum of costs for generating electricity over the lifetime of the generating asset divided by the sum of electrical energy produced over the lifetime of the generating asset. Costs for generating electricity include

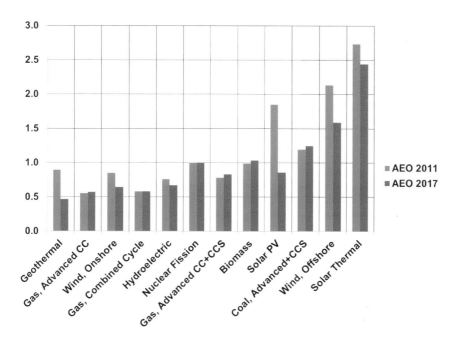

Figure 5-3. Relative Cost of Electricity Generation Normalized to Nuclear Fission.

investment expenditures, operations and maintenance expenditures, and fuel expenditures. Normalization of results to the nuclear fission base line makes it possible to compare the relative costs of generating electricity using different technologies and shows how cost has changed over a period of time.

Using nuclear fission as our baseline, we see in Figure 5-3 that natural gas, hydroelectric, geothermal, onshore wind, and solar PV (photovoltaic) cells are less expensive than nuclear fission in the more recent 2017 study using 2016$ [US AEO, 2017]. No subsidies are included in this comparison. It is worth noting that the cost of solar PV has changed more than any other technology shown in the comparison. The reduction in solar PV cost relative to other technologies increases the commercial competitiveness of solar PV.

Fossil fuels dominate the modern energy infrastructure. Two of their most important functions are the generation of electricity and fuel transportation. Yet fossil fuels exist in finite amounts and are nonrenewable. Peak oil production may not be as imminent as many people once thought, but oil and gas consumption may peak before peak oil production occurs. Until recently, peak oil referred to the supply of oil. Growth of the renewable energy sector, has raised the possibility that peak oil consumption will occur when peak oil demand occurs. In this scenario, the demand for oil will peak as demand for alternative energy sources grows. The replacement of vehicles using internal combustion engines by electric vehicles has raised concerns in the oil and gas industry [Parshall, 2017].

The eventual decline in fossil fuel use will have a significant impact on society. Another possible impact of combustible fossil fuel consumption is climate change, which we now consider in more detail.

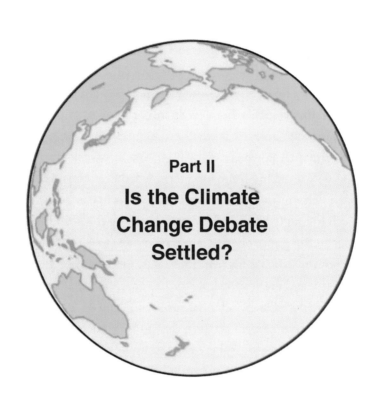

Part II
Is the Climate Change Debate Settled?

6. Is our climate changing?

Many people believe that our current reliance on the combustion of fossil fuels is having an adverse impact on climate. The term "climate change" is widely used to characterize this impact. Our purpose here is to review and evaluate the evidence for climate change. We then consider what might happen if the byproducts of combustion are altering the climate. A plan for making the transition to a sustainable energy portfolio that provides cost-effective energy while protecting the environment is discussed in later sections.

Climate may be defined as the long term weather pattern. Rain, droughts, and storms such as hurricanes are common examples of weather changes. Everyone agrees that weather changes when seasons change, droughts occur, and storms pass through. When we ask "Is our climate changing?" we are asking if the long term weather pattern is changing.

What is changing the climate? Many scientists believe human activity is changing the atmosphere by increasing the amount of greenhouse gases in the atmosphere. Carbon dioxide (CO_2), a by-product of fossil fuel combustion, is a greenhouse gas. Greenhouse gases absorb thermal energy and are responsible for the greenhouse effect, which we discuss in more detail later. Scientists point out that Mars and Venus are illustrations of what carbon dioxide can do to planetary atmospheres. The very cold Martian atmosphere is an example of an atmosphere with too little carbon dioxide and a negligible greenhouse effect. The hot Venusian atmosphere is an example of an atmosphere with so much carbon dioxide that the resulting greenhouse effect substantially increases the temperature of the atmosphere.

The debate about the changing climate is focused on the impact of human activity on changes to the long term weather pattern. Here we use the term anthropogenic climate change to describe climate change due to human activity. Advocates of anthropogenic climate change (ACC) believe that human activity is altering the temperature of the atmosphere, increasing the rate of glacier melting, causing sea levels to rise, and increasing the acidity of ocean water. We discuss these issues below.

7. What is affecting the atmosphere?

Life has been known to affect the atmosphere. We can see this by examining how the content of oxygen in the earth's atmosphere has changed from the origin of the earth approximately 4.5 billion years ago to the present. The atmosphere of the earth has changed during that period of time. Scientists believe that oxygen was not present in the atmosphere when the earth formed. Holland [2006] said that oxygen first appeared in the atmosphere approximately 2.5 billion years ago when cyanobacteria appeared. The two lines in Figure 7-1 depict the range of oxygen partial pressure in the atmosphere. The top line is the maximum value and the bottom line is the minimum value. Oxygen partial pressure is a measure of the amount of oxygen in a gas mixture. The amount of oxygen began to increase as life evolved from relatively simple cyanobacteria into more complex eukaryotes and algae. Today, air typically consists of approximately 21 volume % oxygen and 78 volume % nitrogen. The remaining 1 volume % of air includes argon, carbon dioxide, methane, and water vapor.

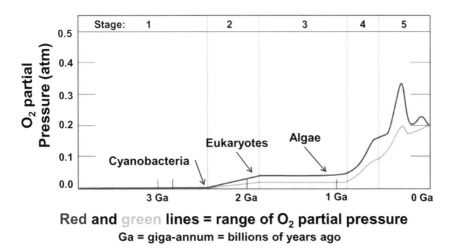

Figure 7-1. Life Affects Oxygen Content in the Atmosphere. [After Holland, 2006]

One source of atmospheric oxygen is plant photosynthesis. Carbon dioxide and water react in the presence of sunlight and chlorophyll to form a carbohydrate and oxygen. Plant photosynthesis converts carbon dioxide into oxygen in the atmosphere. Another photosynthetic process, bacterial photosynthesis, forms a carbohydrate and water. If the water is released into the atmosphere as a vapor, it becomes a greenhouse gas. Scientists have observed that the harmful effects of an increase in atmospheric carbon dioxide are being mitigated by "global greening," that is, an increase in use of atmospheric carbon dioxide by terrestrial vegetation [Keenan, et al., 2016].

Human breathing is the source of a significant amount of carbon dioxide. We inhale air containing 78 volume percent nitrogen and 21 volume percent oxygen. When we exhale, we exhale 78 volume percent nitrogen, 14 to 16 volume percent oxygen, and 4 to 5 volume percent carbon dioxide. We rely on plants to convert exhaled carbon dioxide into oxygen.

The observation that humans are a source of atmospheric carbon dioxide gains in importance when we look at the global population trend. Figure 7-2 shows three global population forecasts from 2010 to 2100 AD are presented. Actual global population reached 7 billion people in 2012. The "High" growth forecast shows that global population may grow to 16 billion by 2100 AD. By contrast, the "Low" growth forecast shows that the population will peak and then decline. A decline in population can occur when more people around the world are educated about the perils of unlimited population growth.

The recent history of population growth in some Western nations suggests that it is possible to reduce population by educating people about birth control. A study of fertility rate shows that the total number of births per woman declines as a country's productivity (measured by GDP) increases. If global population increases, which is likely in the near-term, the associated human exhalation of carbon dioxide will also increase.

Forecasts of global population can be used to determine how much energy will be needed in the future. Figure 3-1 shows the relationship

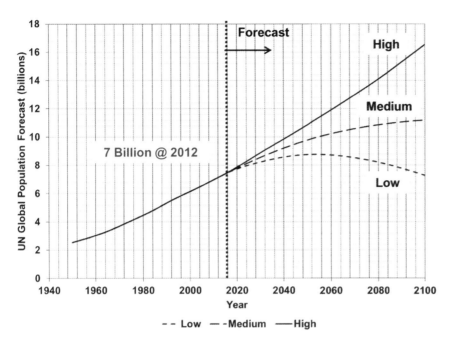

Figure 7-2. UN Global Population Estimates from 1950 to 2100 CE Forecasts Begin at Year 2015. [UN DESA, 2017]

between energy consumption per capita and quality of life as measured by the UN Human Development Index (HDI). The UN HDI is based on life expectancy, education level, and living standard. Western European countries, the USA, Canada, and Australia are examples of countries with a relatively high UN HDI and relatively high per capita primary energy consumption. Countries in sub-Saharan Africa are examples of countries which tend to have relatively low UN HDI and relatively low per capita primary energy consumption.

The relationship between UN HDI and per capita energy consumption presented in Figure 3-2 shows that quality of life depends on energy consumption. According to the figure, one way to improve quality of life in a country is to make more energy available for consumption. A forecast of population combined with a desired quality of life can be used to estimate the amount of energy that must be provided by a country. Improvements in energy conservation may reduce the energy requirement, but it is reasonable to expect that a growing global population and a desire for an improved quality of life in less developed countries will continue to increase demand for energy. Hofmeister observed that [Hofmeister, 2010, page 72] "The world's population is growing; its demand for industrial production and electricity and transportation increases every year." What will be the source of that energy?

One way to minimize the impact of carbon in the atmosphere is to decarbonize our energy consumption and move towards a low-carbon economy. Wind and solar energy are sources of renewable energy that do not require combustion. Figure 7-3 shows three different energy types and their contribution to historical global energy production from 1970 to 2015.

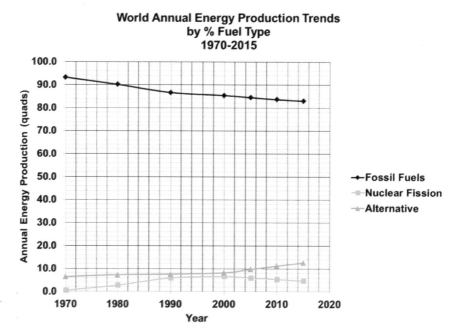

Figure 7-3. Percent Contribution of Different Energy Types to World Historical Energy Production from 1970 to 2015. [US IEO, 2018]

The decline in fossil fuel production shown in Figure 7-3 was greater between 1970 and 1990 than it was between 1990 and 2015. Nuclear fission energy increased between 1970 and 1990, and was relatively flat until the Fukushima disaster in Japan in March 2011. The disaster was caused by an offshore earthquake that generated a tsunami. The tsunami flooded coastal nuclear reactors and lead to the release of radioactive material. There appears to be a downturn in nuclear fission energy production since the Fukushima disaster. The nuclear industry and government regulatory agencies have been reviewing and modifying rules governing the nuclear reactor industry to better prepare for natural disasters, but it is possible that public sentiment against the nuclear fission industry will be long-lived.

The alternative curve in Figure 7-3 includes hydropower and biofuels, in addition to wind and solar energy. Renewable energy production increased over the past decade as more wind farms and solar facilities were built. Wind has been more cost effective than solar, but technological developments are beginning to significantly lower the cost of generating electricity from solar energy.

Biofuels are renewable fuels, but are not as attractive as wind and solar energy because they require combustion. The combustion of carbon based fuels generates heat, carbon dioxide and water vapor. Carbon dioxide and water vapor are both greenhouse gases. Therefore the burning of biofuels generates greenhouse gases in much the same way as the burning of fossil fuels. The idea that biofuels are a desirable replacement for fossil fuels is incorrect if you want to eliminate the emission of greenhouse gases.

8. Are we changing the atmosphere?

One concern about climate change is that the atmosphere is warming. Figure 8-1 shows temperature change from 1850 to 2015. Temperature increased by approximately 1.2°C during that period. This sounds like a relatively small change and we discuss later whether or not it is a significant temperature change. The increase in atmospheric temperature since 1850 coincides with the use of fossil fuels during the industrial age as well as an increase in global human population from approximately 1 billion people to over 7 billion people.

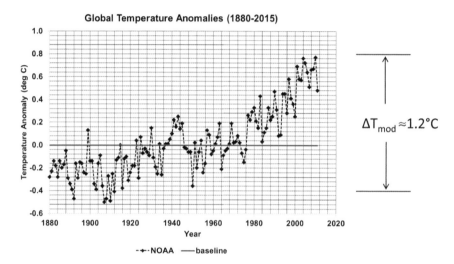

Figure 8-1. Temperature Change from 1850 to 2015. [After Fanchi and Fanchi, 2017]

Coal combustion was used as the primary energy source in industrialized societies prior to 1850. Improvements in drilling technology, and the discovery of oil fields that could provide large volumes of oil at high flow rates made oil less expensive then coal and whale oil. Low cost and the invention of the internal combustion engine helped motivate the conversion from coal to oil as the primary energy source in industrialized societies by the beginning of the 20th century.

One possible explanation for the observed increase in atmospheric temperature is that the sun caused the temperature increase. Figure 8-2 compares global surface temperature with solar insolation between 1978 and 2009. Solar insolation is the solar power arriving at the top of the

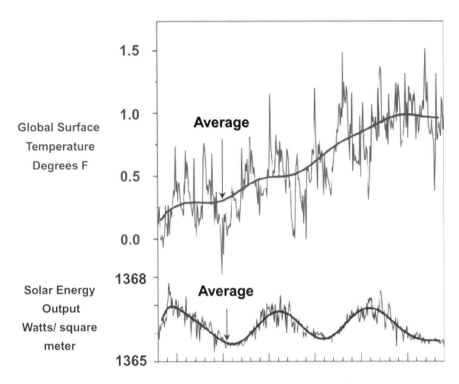

Figure 8-2. Did the Sun cause the temperature increase? [US NOAA Solar Insolation, 2018]

atmosphere per square meter of surface area of atmosphere. It is labelled "Sun's Energy Output" in the figure. We see that the energy arriving at the top of the atmosphere from the sun is following its normal 11-year cycle while global surface temperature has been increasing during the period. The increase in global surface temperature has also abated since the year 2000. Thus, the correlation between solar insolation and global surface temperature is relatively weak.

The explanation for increases in atmospheric temperature that is most widely embraced by mainstream scientists is the greenhouse effect. We have mentioned the greenhouse effect before. Here we define it more carefully.

The greenhouse effect depends on gas molecules in the atmosphere that absorb infrared radiation. Frequencies associated with infrared radiation are part of the electromagnetic spectrum shown in Figure 8-3. Visible light represents a relatively narrow frequency band. Light from the sun covers a wide range of frequencies that includes ultraviolet and infrared frequencies.

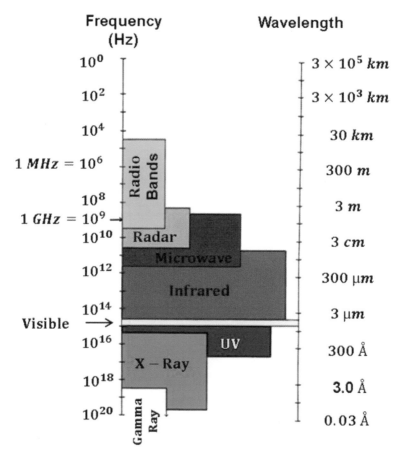

Figure 8-3. The Electromagnetic Spectrum.

Figure 8-4 illustrates the greenhouse effect. Sunlight is energy traveling from the sun to the earth. Some sunlight gets reflected back into space by the atmosphere while the rest enters the atmosphere. The amount of sunlight that reaches the surface of the earth depends on losses in the atmosphere.

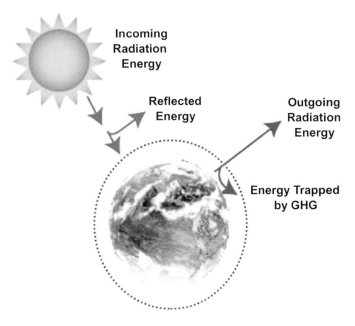

Figure 8-4.   What is the greenhouse effect? [Fanchi and Fanchi, 2017]

Figure 8-5 shows the spectrum of solar irradiance. Solar irradiance is the solar power per unit area of surface area. The vertical axis shows spectral irradiance, or solar irradiance per nanometer (nm) wavelength. The horizontal axis is wavelength of sunlight (in nm). The solid black curve starting at 250 nm, peaking at about 500 nm, and continuing to 2500 nm is the curve for the spectrum of a blackbody radiating at 5250°C. The sun radiates into space like a black body at 5250°C. The solid black curve is the spectral irradiance that would be measured if some of the light was not being absorbed in the atmosphere. The gaps between the blackbody curve and the curve representing radiation at sea level shows that much of the solar energy arriving at the top of the atmosphere is not making it to the surface of the earth. Many of the gaps are associated with light absorption by a greenhouse gas molecule

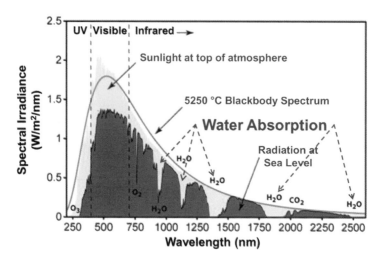

Figure 8-5.   Solar Radiation Losses in Atmosphere. [Solar Spectrum, 2018]

such as carbon dioxide and water vapor. Several of the light absorption gaps are in the infrared.

A portion of the sunlight that arrives at the earth's surface can be transformed into thermal energy and re-radiated back toward space. When that happens, the frequency of incident sunlight can change from visible or ultraviolet to infrared. Re-radiated infrared light has to pass through the atmosphere. Some gas molecules can absorb infrared light. If these molecules are in the atmosphere, they behave as greenhouse gases. When a greenhouse gas molecule absorbs infrared light, the increase in kinetic energy of the gas molecule increases the temperature of the gas in the region of the molecule. The resulting increase in atmospheric temperature is the greenhouse effect.

The most common greenhouse gases are water vapor, carbon dioxide, methane, and nitrous oxide. Other greenhouses gases include volatile organic compounds such as hydrofluorocarbons. There is more carbon dioxide than methane in the atmosphere. Greenhouse gases are the focus of attention in atmospheric warming.

One way to measure the concentration of carbon dioxide in the atmosphere is represented by the Keeling Curve. Charles David Keeling began measuring carbon dioxide concentration in the atmosphere in 1958. He made his measurements at the Mauna Loa Observatory on the Big Island of Hawaii. The top of the Mauna Loa volcano is a good location for measuring the composition of gases in the atmosphere because it is above typical pollution sources closer to sea level. Figure 4-1 shows that there has been a steady increase in carbon dioxide concentration since measurements began in 1958. The sawtooth structure of the curve shows an annual cycle. The carbon dioxide concentration was originally measured at a little over 310 parts per million. Today it is approximately 400 parts per million. These measurements show that carbon dioxide concentration in the atmosphere has been increasing since the middle of the 20th century.

We observed in Figure 8-1 that the global temperature change from 1850 to 2015 is about 1.2°C. Figure 8-6 shows the methane and carbon

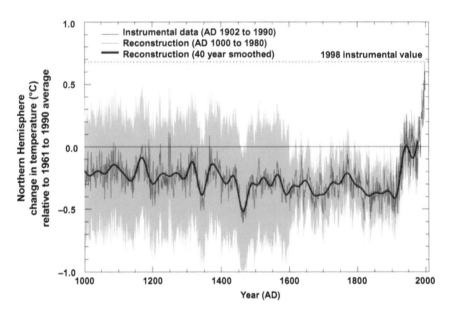

Figure 8-6. History of Atmospheric CO_2 and CH_4 from 1000 AD to 2000 AD. [IPCC, 2001, Figure 5, page 29]

36 | The Goldilocks Policy

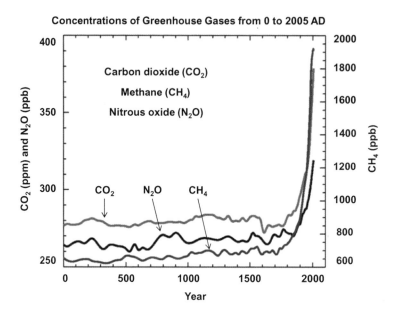

Figure 8-7. Atmospheric CO_2, CH_4, and N_2O — 2,000 Years of History. [IPCC, 2007]

dioxide concentrations in the atmosphere for the period of time ranging from about 1000 AD to 2000 AD. The shapes of these curves look like hockey sticks with the blades pointing up. Both methane and carbon dioxide are greenhouse gases.

Figure 8-7 shows the concentrations of carbon dioxide, methane and nitrous oxide over a 2000 year period. If we look back over the past 2000 years, we see that these greenhouse gas concentrations were relatively constant until roughly the beginning of the industrial revolution in the 18th century. The onset of significant increases in greenhouse gas concentrations when human beings began to use combustible fuels on a grand scale is part of the reason that many people argue that human activity is the source of the increase in greenhouse gas concentration in the atmosphere and a source of global warming.

The increase in greenhouse gas concentration is a source of concern if the measurements are correct. How do we measure methane and carbon dioxide concentration going back over 1000 years?

An important method for measuring the composition of the earth's atmosphere in the distant past is ice coring in Antarctica. Ice cores have been obtained at Lake Vostok, Antarctica by drilling into the ice with a core drilling bit and extracting cores of ice from different depths. Antarctic ice has presumably not melted once it was formed. The formation of ice by compaction of snow or freezing water would capture samples of air that was present at the time the ice was formed. The depth of the ice sample represents a point of time in the earth's past. Deeper ice samples are older than shallow ice samples. When a sample of ice is brought to the surface and allowed to melt, the gas released from pores in the ice core represents air present at the time the ice was formed. A sample of the released gas is captured and analyzed. The composition of the gas that was trapped in the ice provides information about the composition of the atmosphere at the time the ice was formed.

Vostok ice core data [Petit *et al.*, 2001] were used to estimate atmospheric temperature, methane and carbon dioxide concentration, and dust concentration in the past. Higher dust levels are thought to be caused by cold, dry periods. Figure 8-8 shows the Vostok temperature change and two greenhouse gases (GHG). Methane is reported in parts per billion, and carbon dioxide is reported in parts per million. There appears to be a correlation between GHG and temperature change dating back over 800,000 years.

The relationship between atmospheric carbon dioxide concentration and temperature change from Antarctic ice cores is displayed in Figure 8-8. The curves represent 800,000 years of history. Modern measurements of carbon dioxide are shown as the nearly vertical dark line on the right hand side of the figure. Historically, Vostok carbon dioxide concentration data peaked around 300 parts per million. Today, Mauna Loa data indicate that atmospheric carbon dioxide concentration is approximately 400 ppm, which is significantly greater than the Vostok historical peaks.

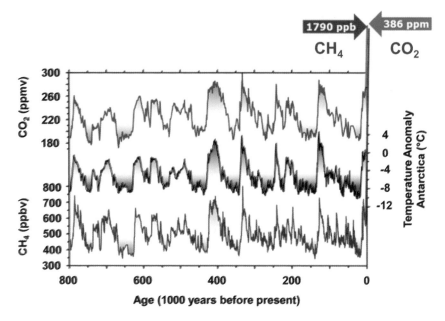

Figure 8-8. Historical Vostok Temperature Change and Composition. [Vostok Temperature and GHG, 2018]

An apparent correlation between temperature and atmospheric carbon dioxide concentration is displayed in Figure 8-8 for the past 800,000 years. Temperature anomaly is shown in the center curve. Temperature anomalies are not yet exceeding historical values. The concern is that the temperature anomaly will increase to dangerous levels as a consequence of increased atmospheric carbon dioxide concentration.

Figure 8-9 shows forecasts of greenhouse gas emissions and temperature change from the US Global Change Research Project. The forecasts show that the temperature increase can range from approximately 4°F (≈2.2°C) at the "lower emission" forecast to 8°F (≈4.4°C) at the "even higher emission" forecast.

It is possible that we can minimize the temperature increase by reducing the amount of carbon dioxide being emitted into the atmosphere. A U.S. Department of Energy study [U.S. DoE Sequestration,

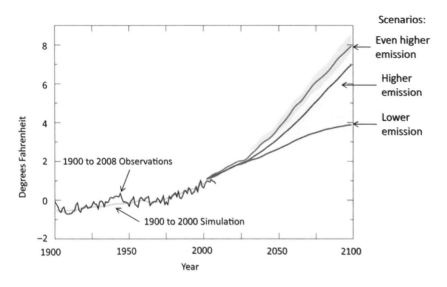

Figure 8-9.   Effect of Emissions on ΔT. [US GCRP, 2009]

1999] reported that 550 parts per million would be a stable concentration of atmospheric carbon dioxide. The pre-industrial concentration was 288 parts per million and the current value measured at Mauna Loa is approximately 400 parts per million. The stable value of 550 ppm would actually be higher than the current value and the pre-industrial value. Global greening, carbon capture and storage (CCS), and carbon capture, utilization and storage (CCUS) are mechanisms that can slow the increase in concentration of atmospheric carbon dioxide. See Section 11 for more discussion.

The concern that temperature will increase as a consequence of increased atmospheric carbon dioxide concentration is based on the assumption that an increase in carbon dioxide concentration will drive an increase in temperature. Figure 8-10 presents approximately 450,000 years of Vostok data. The temperature change is displayed on the left hand axis and the carbon dioxide concentration in parts per million is displayed on the right-hand axis. A cursory look at approximately 450,000 years of Vostok data shows an apparent correlation between the curves.

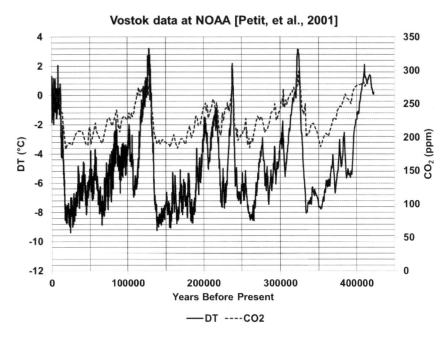

Figure 8-10. Does temperature increase occur before CO_2 increase? [Petit et al., 2001]

A closer look at the data is shown in Figure 8-11. The black curve labeled Delta T is the change in temperature, and the lighter (red) curve shows carbon dioxide concentration in parts per million. Figure 8-11(a) highlights data from 100,000 to 150,000 years ago. Figure 8-11(b) highlights data from 200,000 to 250,000 years ago. The dotted circles focus on the beginning of the increase in temperature and the beginning of the increase in carbon dioxide concentration. Does the data show that carbon dioxide concentration increases before temperature increases? This leads to a question raised by anthropogenic climate change skeptics: does an increase in carbon dioxide concentration drive the increase in temperature or does the increase in temperature drive the increase in carbon dioxide concentration? Another possibility is that an unknown factor is driving both the change in temperature and carbon dioxide concentration.

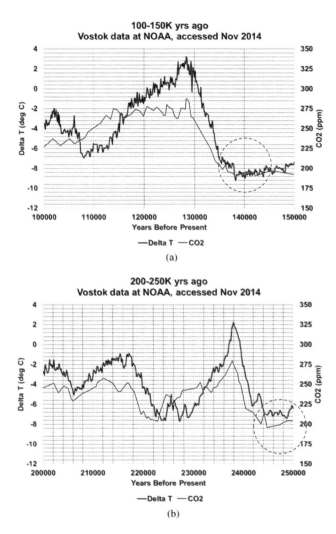

Figure 8-11. (a) Expanded View of Vostok data: 100 – 150K years ago [Petit, *et al.*, 2001], (b) Expanded View of Vostok data: 200 – 250K years ago, [Petit, *et al.*, 2001]

9. Has climate change occurred before? The long view

At this point our assessment of the data shows that there is an increase in atmospheric carbon dioxide concentration that is significantly above historical levels covering the last 450,000 years or so. There appears to be a correlation between changes in temperature and greenhouse gas concentration. The greenhouse effect is the most likely mechanism for the correlation, but there could be other mechanisms involved. We have also noticed that the temperature increase since 1850 is approximately 1.2°C, which is less than temperature variations observed from analysis of Lake Vostok ice core. We now consider the change in climate over an even longer period of time.

We can understand long term climate changes on the earth such as ice ages by considering Milankovitch cycles. Milutin Milankovitch, a Serbian geophysicist and climatologist (1879–1958), provided the modern explanation of long term climate changes. He used astronomical mechanisms based on the position of the earth relative to the sun.

Milankovitch recognized three cycles associated with the motion of the earth. The earth moves like a top spinning around the sun. The longest cycle lasts approximately 100,000 years and is called the eccentricity cycle. Eccentricity refers to the earth's position relative to the sun as a consequence of its elliptical orbit (Figure 9-1). The distance between the sun and the earth changes over time. The next cycle is approximately 41,000 years long and is called the obliquity cycle or tilt. It is due to

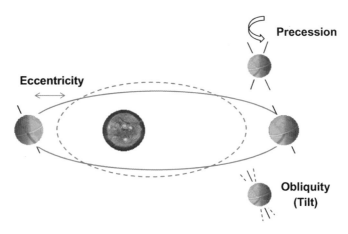

Figure 9-1.   Milankovitch Cycles.

the variation of the tilt of the earth from approximately 21.5° to 24.5° relative to an axis that is perpendicular to the orbital plane of the earth. A precession or wobble occurs in conjunction with the rotation of the tilted earth. The duration of the precession cycle is approximately 23,000 years. Table 9-1 summarizes the duration of the cycles. Milankovitch cycles have been used to improve the accuracy of the geologic time scale used by geologists.

Table 9-1. Duration of Milankovitch Cycles.

Cycle	Duration (yrs)
Eccentricity (distance)	100,000
Obliquity (tilt)	41,000
Precession (wobble)	23,000

Annual seasons are related to the variation of the sunshine impacting different parts of the earth each year. The axial tilt has the greatest effect on seasons, followed by the variation of distance from the earth to the sun associated with the eccentricity of the orbit. Other planets, especially Jupiter, and the albedo (reflectivity) of the earth effect the seasons, but to a lesser extent than axial tilt and eccentricity.

In addition to driving annual seasons, Milankovitch cycles drive ice age cycles. The principal cause of ice ages is the eccentricity cycle. Axial tilt and precession are secondary effects.

Figure 9-2 and Table 9-2 show the timeline for five major ice ages during the past 2.4 billion years. We are currently in the ice age called the Current Ice Age or Quaternary Glaciation. The variation of average global temperature of the ice ages has been approximately 14°C. For comparison, the temperature increase we are seeing today is a little greater than 1°C, which is much less than the temperature variation seen during the five major ice ages.

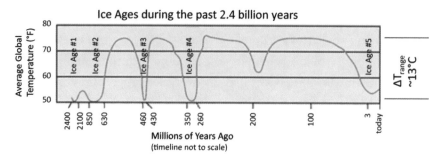

Figure 9-2. Five Major Ice Ages. [UGS Ice Ages, 2018]

Table 9-2. Major Ice Ages.

#	Ice Age	MYA
5	Current Ice Age (aka Quaternary Glaciation)	~ 3 to now
4	Karoo Ice Age	~ 350 to 260
3	Andean-Saharan glaciation	~ 460 to 430
2	Sturtian/Marinoan glaciation	~ 850 to 630
1	Huronian glaciation	~ 2400 to 2100

Figure 9-3 shows that temperature change varies within an ice age. The temperature variation corresponds to glacial and interglacial periods. Here we see the temperature change during the past 450,000 years has varied by approximately 9°C.

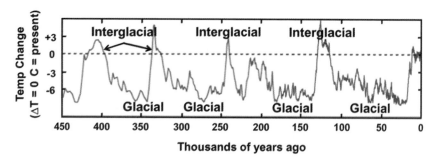

Figure 9-3. Glacials and Interglacials in the Past 450,000 Years. [UGS Ice Ages, 2018]

If we look at near surface temperatures in the northern hemisphere over the past 11,000 years (Figure 9-4), we see that we are now in an interglacial period of the current ice age. The temperature change over the last 11,000 years has been approximately 6°C. The modern temperature change has been about 1°C, maybe 2°C if we go back to the coolest days of the Little Ice Age (ca 1500–1600). There have been periods of warming in the past like we are in today, such as the Medieval Warm Period. Many cultural changes occurred during the Medieval Warm Period, which is also known as the Medieval Climate Anomaly.

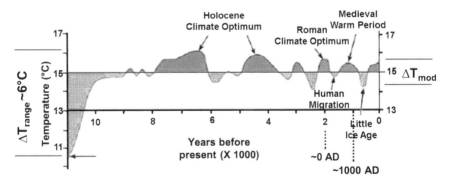

Figure 9-4. Climate Swings During Past 11,000 Years. [Archibald, 2018]

Some authors [Sussman, 2010, pages 26–30 and footnote 58] have pointed out that the Medieval Warm Period (MWP) may have created a climate that made it possible for the Vikings to discover Iceland, Greenland and Newfoundland. The MWP occurred during the 400-year period from 900 to 1300 A.D. Viking settlements were established in these regions during the MWP. It appears that a mild warming in the northern hemisphere facilitated exploration.

The existence of the MWP has been a source of dispute with some researchers. The 2001 IPCC Third Assessment Report contained a plot of temperature change versus time attributed to Mann *et al.* [1998, 1999]. They determined the average temperature of the Northern Hemisphere from 1961 to 1990. They then plotted departures in temperature from the 1961 to 1990 average for the period ranging from 1000 A.D. to 1990 A.D. The plot showed a relatively constant temperature from 1000 A.D. until approximately 1900 A.D. At that point, the temperature departure from the 1961 to 1990 average increased sharply. The resulting plot looked like a hockey stick with the handle flat and the edge up. The temperature change associated with the MWP from 950 A.D. to 1250 A.D. and the Little Ice Age from 1450 A.D. to 1850 A.D. were no longer present.

Critics have argued that the Mann *et al.* [1998, 1999] plot was based on an inaccurate data set. For example, the plot included temperature estimates from tree rings, corals, ice cores, and historical records. The measurements included temperatures recorded in urban areas where the temperatures were elevated by the Urban Heat Island effect. Satellite data were not included in the study.

McIntyre and McKitrick [2003] published a correction to the Mann *et al.* data. Mann *et al.* [2004] published their own correction in the journal *Nature* in 2004. McIntyre and McKitrick sought access to more data so they could provide a comprehensive analysis of temperature departures from 1000 A.D. to 2000 A.D. The journal *Nature* did not publish their findings. McKitrick [2005] gave his perspective of the dispute and expressed concern about bias appearing in the IPCC Reports. Both the MWP (also known as Medieval Climate Anomaly or MCA) and Little Ice Age (LIA) are displayed in Figure 5.8 of the 2013 IPCC Report [Masson-Delmotte *et al.*, 2013].

10. What might be the consequences of anthropogenic climate change?

Many influential people who believe that anthropogenic climate change is occurring want society to immediately transition to a low-carbon economy. The cost to convert from a society that depends on carbon-based fuels to a low-carbon economy is enormous. Is the immediate transition necessary? Let us consider some of the consequences of adverse climate change if we do not undertake an immediate transition.

Figure 10-1 shows that the global ice sheet has been shrinking in the Arctic for the past 18,000 years. The shrinking of the Arctic ice sheet is not a modern phenomenon. By contrast, scientists have observed growth of the ice volume in Antarctica. The change in ice volume in Antarctica partially offset the reduction in Arctic ice volume. A rise in sea level would result in the inland movement of shoreline and flooding of coastal areas.

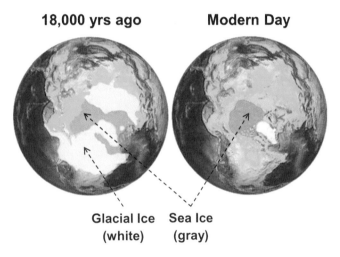

Figure 10-1. Arctic Ice Sheet Shrinking from 18,000 Years Ago to Present. [US NOAA Arctic Ice Sheet, 2018]

By contrast, Smil considered fossil fuel consumption and its possible impact on sea level rise in Antarctica [Smil, 2015, page 440]. Referring to the troposphere, which is the lowest level of the atmosphere that extends from the earth's surface to a height of approximately 10 km, Smil suggested that "the eventual exhaustion of fossil energies is most unlikely because the burning of coal and hydrocarbons is the principal source of anthropogenic CO_2 and the combustion of available fossil fuel resources would raise the tropospheric temperature high enough to eliminate the entire Antarctic ice sheet and cause a sea-level rise of about 58 m." [Smil, 2015, page 440] The estimate of sea-level rise was presented in a 2015 study by Winkelmann *et al.* who pointed out that it would take 10,000 years for the sea-level to rise 58 m (190 ft) [Winkelmann *et al.*, 2015]. Winkelmann *et al.* said that projected sea-level rise during this century would be in the range from −6 cm (−0.20 ft) to 14 cm (0.46 ft) depending on the amount of anthropogenic CO_2 emitted into the atmosphere. If atmospheric temperature is low enough, corresponding to low anthropogenic CO_2 emission levels, more ice could form in Antarctica and sea level would fall, corresponding to the negative sea level rise. One way to interpret the Winkelmann *et al.* study is to recognize that there is some time before sea level rise leads to catastrophic flooding.

Average sea level worldwide has been increasing at approximately 0.6 ft per century [US EPA Sea Level, 2014]. Anthropogenic climate change advocates believe that the rate of sea level rise could triple by 2,100 if greenhouse gas emissions continue to be added to the environment. One way to put sea level rise in perspective is to note that storm surge from a hurricane can be 20 feet higher than sea level [US NOAA NHC, 2014]. In the case of storm surge, flooding will usually recede within days. Sea level rise, on the other hand, will require adaptation to a long-term change in sea level.

11. How can we reduce greenhouse gas emissions?

One way to minimize greenhouse gas emissions is to inject the gases into geologic formations in a process called geologic sequestration. Coalbeds can be used to store carbon dioxide. Carbon dioxide injection into coal displaces methane because carbon dioxide is preferentially absorbed by coal when compared to methane absorption. Carbon dioxide sequestration in coal can increase the production of methane for commercial use. The combustion of methane yields more carbon dioxide which can then be injected into the coalbed. Sequestration of carbon dioxide in coal is a plausible method for the long-term storage of some carbon dioxide.

Other sequestration techniques use known hydrocarbon reservoirs for storage. The injected greenhouse gases can enhance hydrocarbon recovery, but care must be taken to keep reservoir pressure less than fracture pressure. If injection pressure exceeds fracture pressure, the reservoir rock can be fractured so fluids can flow out of the reservoir. Another geologic storage location is a salt dome. A cavern can be formed in the salt dome by injecting water into the dome. The cavern becomes an impermeable storage tank.

Geologic sequestration must have enough volume to take all of the carbon dioxide that will be emitted into the atmosphere. Some scientists argue that there is not enough geologic storage capacity to store all of the greenhouse gases produced globally over an extended period of time.

There are alternatives to geologic storage of greenhouse gases. Some have suggested putting greenhouse gases into ocean locations where the mixture of water and gas at low temperature and high pressure can form a hydrate. Hydrates are gas molecules bounded by water molecules in ball-like structures. The problem with the idea of ocean hydrate storage is that the hydrate structures can break and release the bound gas molecule if the oceans warm up. Consequently, greenhouse gas in the hydrates would be released into the atmosphere and contribute to the greenhouse effect.

Another alternative to geologic sequestration is to change the economy from a fossil fuel-based economy to a low-carbon economy. Most countries in the European Union are dependent on energy imports. A few have viable fossil fuel industries and France relies on nuclear fission energy. Geopolitical events, such as the Russian annexation of Crimea, unrest in Ukraine, and political instability in the Middle East, have motivated a sense of urgency to transition to sustainable energy.

The EU has made a decision to build an infrastructure that could support a low-carbon economy. It will include renewable energy sources such as wind and solar. One plan relies largely on coastal wind farms, and solar power plants throughout the Mediterranean, northern Africa, and the Middle East. The EU plan is to build an electrical grid to distribute energy throughout the region. The goal is to have the new energy system in place by 2050. Support for the plan is motivated by the desire to achieve a secure energy supply, an environmentally sustainable society, and a competitive economy.

The USA and China signed an agreement in November 2014 to reduce greenhouse gas emissions. The USA is expected to reduce emissions by up to 28% below its 2005 levels by the year 2025. In exchange, China will set the peak of its emissions by 2030 or earlier. They will not have to reduce their emissions before 2030, although China agreed to increase their share of energy consumption from non-fossil fuels to about 20% of their total consumption by 2030.

The information displayed in Figure 11-1 can help us understand the rationale for reducing greenhouse gas emissions. Tonnes of carbon dioxide emitted per person as a function of time are presented in the figure. A tonne

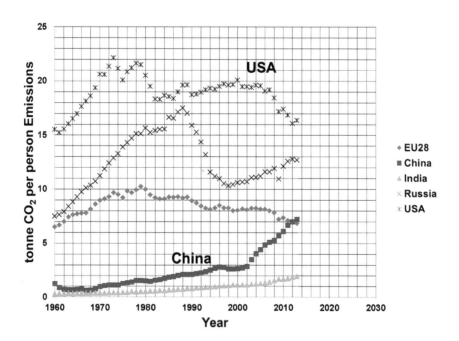

Figure 11-1. Selected 1960–2013 Per Capita Emissions by Country. [Boden et al., 2015]

is a metric ton, or 1000 kg, that equals approximately 2204.6 pounds. The USA is at the top of this chart while China is relatively low, even though there has been a substantial increase in emissions per person since around 2002. Population estimates for the period from 2012 to 2014 show that the USA has approximately 310 million people and China has approximately 1,339 million people. The ranking changes when we consider total emissions for each country, rather than per capita emissions.

Figure 11-2 shows total megatonnes (million tonnes) of carbon dioxide emissions. In this case, China is emitting more carbon dioxide into the atmosphere than any other country in the world. The USA has dropped into second place. The extrapolated lines show what will happen by 2030 if we assume business as usual for both China and the USA. The USA is expected to reduce its emissions to the 2005 level indicated in the figure. The dashed line from 2005 to 2025 is an extrapolation

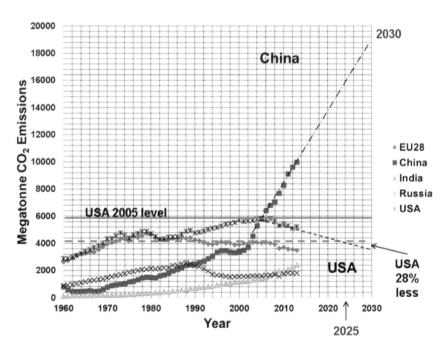

Figure 11-2. 1960–2013 Total Emissions and 2014 USA–China Agreement.

of USA emissions. By 2025, the 2014 USA–China agreement requires the USA to reduce carbon dioxide emissions by at least 28%, which is essentially the same emission reduction that would occur if the existing rate of USA emission reduction continues. Emission reduction in the USA is being achieved by replacing coal with natural gas in power plants, closing coal-fired power plants, and reducing emissions from vehicles. The figure also shows a linear extrapolation of Chinese emissions to 2030. China does not have to reduce emissions until 2030, so the extrapolated value is an estimate of the peak at the end of the agreement. If China continues business as usual, Chinese emissions could be almost twice the emissions they are currently today. China is free to reduce or increase the rate of emissions during the duration of the 2014 USA–China agreement. An increase in the rate of emissions is conceivable if China decides to increase the rate at which it is installing fossil-fuel fired power plants, or the fleet of fossil-fuel powered vehicles increases significantly.

12. Is the climate change debate settled?

We have discussed a variety of mechanisms that affect the environment. There is another possibility that should be considered. Figure 12-1 shows world population (in billions), atmospheric temperature change since 1850, and carbon dioxide emissions since 1850. Both temperature change and carbon dioxide emissions are expressed relative to 2008 emissions. The dashed red line shows the end of World War II to indicate a major historical event and a time when technology made significant leaps forward. Many technological advances were made in energy generation, medicine, and food production during wartime that allowed the post-war world to grow both economically and in population rapidly after the war ended. Notice that before WWII, population growth was relatively constant. After WWII, population growth increased substantially. The post-WW II population

Is the Climate Change Debate Settled? | 53

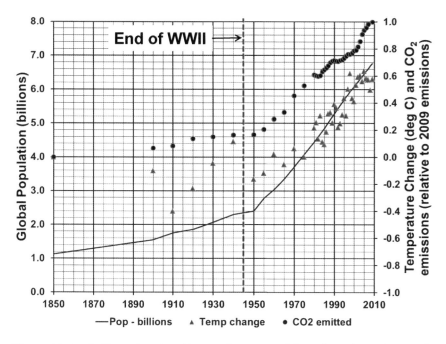

Figure 12-1. Is There Another Human Connection? [Fanchi and Fanchi, 2017]

growth shows the modernization of the industrial world and the growth in the number of people on the planet. Today the world population is over 7 billion people. The increase in the number of people appears to be correlated to increases in temperature change and carbon dioxide emission.

Figure 12-2 shows the correlation between carbon dioxide emission and global population. It also presents the correlation between temperature change and global population. All three variables — population, carbon dioxide emissions, and temperature change — appear to be correlated. We could argue that growth in population drives increases in carbon dioxide emission because people exhale carbon dioxide and consume more carbon-based energy. How should we include population growth in our analysis? Is population growth driving changes in both carbon dioxide emissions and temperature?

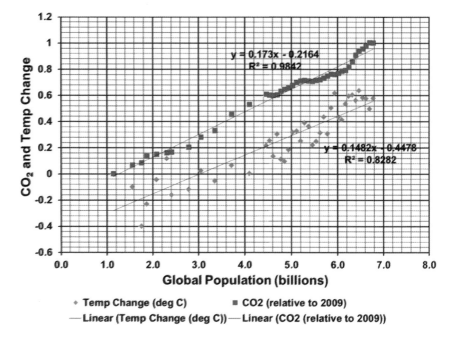

Figure 12-2.   Is There Another Human Connection? [Fanchi and Fanchi, 2017]

12.1 *Population growth*

We seem to have an issue with growing population. Authors have expressed concern about population growth in the past. For example, Thomas R. Malthus presented his theory of population growth in *An Essay on the Principle of Population* [Malthus, 1798]. Malthus was a political economist who saw in nature that plants and animals produced more offspring than can survive. He believed that human population growth would occur at a faster rate than the growth in food supply if left unchecked.

More recently, Paul Ehrlich [1968], with uncredited co-authorship by his wife Anne Ehrlich, wrote in *The Population Bomb* that the earth could not support the expanding population. According to Ehrlich, the competition for resources resulting from overpopulation would lead to mass starvation and major societal upheavals. Ehrlich often worked with John Holdren, who was most widely known as the Director of the Office of Science and Technology Policy during the Obama Administration. In a revisit of *The Population Bomb*, Paul and Anne Ehrlich expressed their pride in their work and concluded that "humanity has reached a dangerous turning point in its domination of the planet" [Ehrlich and Ehrlich, 2009, page 69].

The Ehrlichs and Holdren have argued that government should regulate human reproduction in an attempt to regulate the use of resources. They wrote that "it has been concluded that compulsory population-control laws, even including laws requiring compulsory abortion, could be sustained under the existing Constitution if the population crisis became sufficiently severe to endanger the society" [Ehrlich *et al.*, 1977, page 837]. The concluded that the "most compelling arguments that might be used to justify government regulation of reproduction are based upon the rapid population growth relative to the capacity of environmental and social systems to absorb the associated impacts. To provide a high quality of life for all, there must be fewer people" [Ehrlich *et al.*, 1977, page 837]. The Ehrlichs and Holdren said that a "massive campaign to restore a high-quality environment in North America and to de-develop the United States" must occur so that resources and energy "from frivolous and wasteful uses in overdeveloped countries" could be used to fill "the genuine needs of underdeveloped countries" [Ehrlich *et al.*, 1973, page 279].

12.2 *IPCC versus NIPCC: Is this the debate?*

The mainstream view of climate is provided by the United Nation's Intergovernmental Panel on Climate Change (IPCC). The IPCC was "established by the United Nations Environment Programme (UNEP) and the World Meteorological Organization (WMO) in 1988 to provide the world with a clear scientific view on the current state of knowledge in climate change and its potential environmental and socio-economic impacts. In the same year, the UN General Assembly endorsed the action by WMO and UNEP in jointly establishing the IPCC." [IPCC, 2018] Energy advisor Daniel Yergin said that the establishment of the IPCC in 1988 was the decisive step "that would frame how the world sees climate change today" [Yergin, 2011, page 461.] Yergin said that the IPCC "was not an organization in any familiar sense. Rather it was a self-regulating, self-governing organism, a coordinated network of research scientists who worked across borders, facilitated by cheaper and better communications." [Yergin, 2011, page 461] The purpose of the IPCC was to review and assess "the most recent scientific, technical and socio-economic information produced worldwide relevant to the understanding of climate change. It does not conduct any research nor does it monitor climate related data or parameters." [IPCC, 2018] Bert Bolin, a Swedish meteorologist, was the founding chairman of the IPCC and "was at the center of the growing international climate work" [Yergin, 2011, page 461].

A second, independent group has formed to verify and, if necessary, counter arguments made by the UN sanctioned IPCC. The second group is called the Nongovernmental International Panel on Climate Change (NIPCC) and has the goal of independently evaluating the impact of increasing atmospheric carbon dioxide on the earth's biosphere. Both groups include scientists and engineers with good credentials in science and technology.

Table 12-1 compares the views of the UN IPCC and NIPCC on several different topics. There are topics where there is agreement and there are topics where there is disagreement. For example, both groups agree that climate is changing. The UN IPCC argues that climate change is due to human activity. The NIPCC says that the changing climate is a natural variation of climate.

Several other topics are noted in the table. One more topic to highlight is the validity of global climate models. Yergin identified two technological advances that broadened "the scientific base for

Table 12-1. Contrasting Views of Climate Change by IPCC and NIPCC.

Topic	IPCC UN-2013	NIPCC CCR-II
Climate is changing	Due to human activity	Natural variation
Sea levels	Rate of sea level rising is increasing	Sea level rising in some places, falling in others
Precipitation over land increased in 20th century	Land in Northern Hemisphere	Global average
Cryosphere (melting ice)	Increasing	Global balance
Earth surface temperature increasing	Increasing from 1880 to 2012	Recent cooling has counteracted warming
CO_2 increase due to humans	Due to fossil fuels	Due to growing population
Extreme weather	Increasing	Within natural variation
Global climate models	Trustworthy	Unreliable

understanding climate." [Yergin, 2011, page 445]. Satellites could be used to measure and observe changes in the earth's weather patterns, and computer models could be used to analyze climate data. The UN IPCC says global climate models are trustworthy whereas the NIPCC says that they are unreliable. The NIPCC points out that global climate model forecasts are not reliable because they do not adequately model all of the mechanisms that affect climate behavior. For example, one mechanism that is being studied is the exchange of gas molecules between the ocean and the atmosphere. Attempts to validate the models using historical data have shown the limitations of the models [Hourdin, *et al.*, 2017]. We can conclude that two groups of individuals have come to different conclusions. It is worth noting that governments are continuing to provide substantial funding for the study of climate change, which suggests there are still outstanding issues.

John Hofmeister provided a blunt view of the debate. He spent 25 years in energy consuming companies before joining Shell Oil Company in 1997. He served as president of Shell Oil Company from 2005 to 2008. His career gave him a chance to see the energy industry as a consumer and then a producer. Hofmeister considered the climate change debate and concluded [Hofmeister, 2010, page 64]: "… debating climate change is a fantastic waste of time and human energy. There is no agreement on what it is or isn't. There is no set of measures accurate enough to be credible to present a clear and present danger. There is no rebuttal for the argument that we have always had cycles of global warming and global cooling, and Earth has adjusted accordingly."

Another view was expressed by Steven Koonin who was the Under Secretary for Science in the United States Department of Energy during President Barack Obama's first term. Koonin had prior experience as a physics professor and provost at Caltech. In an opinion piece in the *Wall Street Journal* (Sep. 19, 2014), Koonin said "We are very far from the knowledge needed to make good climate policy." He observed that "the idea that climate science is settled" is misguided. He believes the idea has distorted public and policy debate, and inhibited scientific and policy discussions.

Koonin supported his opinion by pointing out that global climate models provide widely varying forecasts of how climate behaves. For example, Koonin said that global climate models "roughly describe the shrinking extent of Arctic sea ice observed over the past two decades, but they fail to describe the comparable growth of Antarctic sea ice." Koonin also wrote that "natural influences and variability are powerful enough to counteract the present warming influence exerted by human activity."

12.3 *COP21: The Paris climate meeting*

A meeting of almost 200 nations was held in Paris from November 30 to December 11, 2015. The purpose of the meeting was to seek an international agreement to reduce the impact of human activity on the climate. It was convened as COP21, the 21st United Nations Framework Convention on Climate Change (UNFCCC) Conference of Parties.

The COP21 Paris Climate Agreement was adopted by 195 countries on December 12, 2015 [COP21 Agreement, 2015] and is expected to enter into force in 2020. The agreement is "a global action plan to put the world on track to avoid dangerous climate change by limiting global warming to well below 2°C" above pre-industrial levels. Temperature at the pre-industrial level is often taken to mean an average temperature in the 18th century before the beginning of the industrial age.

Key elements of the COP21 Paris Climate Agreement are summarized below [COP21 Agreement, 2015]:

- Governments agreed to reduce emissions with a long-term goal of keeping the increase in global average temperature to well below 2°C (3.6°F) above pre-industrial levels.
- Governments agreed to limit the increase to 1.5°C (2.7°F), since this is expected to significantly reduce risks and the impact of human activity on climate.
- Governments agreed that global emissions need to peak as soon as possible, recognizing that this will take longer for developing countries. Governments are expected to undertake rapid reductions in emissions as soon as possible in accordance with the best available science.
- Governments agreed to be transparent by providing reports to each other and the public.
- Governments agreed to be adaptable. They will strengthen societies' ability to deal with the impact of human activity on climate; and provide continued and enhanced international support for adaptation to developing countries.

Developed countries agreed to help developing countries take action to reduce emissions and build resilience to the impact of human activity on climate. Other countries are encouraged to provide such support voluntarily. Developed countries intend to mobilize US$100 billion per year until 2025 when a new collective goal will be set.

On December 9, 2015, United States Secretary of State John Kerry pointed out the importance of global cooperation while attending the COP21 meeting: "The fact is that even if every American citizen biked to work, carpooled to school, used only solar panels to power their homes, if we each planted a dozen trees, if we somehow eliminated all of our domestic greenhouse gas emissions, guess what — that still wouldn't be enough to offset the carbon pollution coming from the rest of the world.

"If all the industrial nations went down to zero emissions — remember what I just said, all the industrial emissions went down to zero emissions — it wouldn't be enough, not when more than 65 percent of the world's carbon pollution comes from the developing world."

The Trump Administration withdrew the United States from the COP21 Paris Climate Agreement in June 2017 because the agreement was unfair to American workers and taxpayers. The Trump Administration expressed concern that the agreement would have a greater negative impact on the United States economy than a positive impact on the global environment.

12.4 *The oil and gas climate initiative*

How has the oil and gas industry responded to concerns about anthropogenic climate change? Ten members of the oil and gas industry launched the Oil and Gas Climate Initiative (OGCI) in 2014. According to the OGCI website [OGCI Founding, 2018], "OGCI is a voluntary, CEO-led initiative which aims to lead the industry response to climate change." The purpose of OGCI is to "pool expert knowledge and collaborate on action to reduce greenhouse gas emissions." The ten founding companies are BP (Britain), CNPC (China National Petroleum Corporation), Eni (Italy), Equinor (formerly Statoil of Norway), Pemex (Mexico), Petrobras (Brazil), Repsol (Spain), Saudi Aramco (Saudi Arabia), Shell (The Netherlands) and Total (France). None of the founding companies was headquartered in the United States.

The policy and strategy of the OGCI founding members states that "We, the leaders of the ten major oil and gas companies, are committed to the direction set out by the Paris Agreement on climate change. We support its agenda for global action and the need for urgency. Through our collaboration in Oil and Gas Climate Initiative (OGCI), we can be a catalyst for change in our industry and more widely."

"OGCI aims to increase the ambition, speed and scale of the initiatives we undertake as individual companies to reduce the greenhouse gas footprint of our core oil and gas business — and to explore new businesses and technologies." It is important to note that OGCI is not abandoning its core oil and gas business; instead, it is focusing on mitigating the negative effects of consuming combustible fuels.

OGCI announced in September 2018 that three major oil companies headquartered in the United States had joined the ten founding members of OGCI [OGCI US, 2018]. The three new member companies were Chevron, ExxonMobil, and Occidental Petroleum.

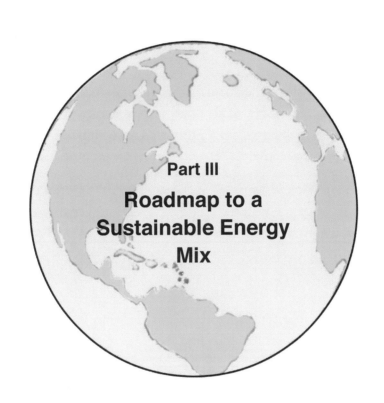

13. Trend toward decarbonization

The industrial age used coal as its fuel of choice. Coal replaced wood and helped reduce deforestation in nations like Britain. Oil replaced coal and whale oil. Today we are dependent on oil and observing a move toward natural gas, which is predominantly methane. The amount of carbon in each molecule of carbon-based fuel has been declining as we move from coal to oil to gas. The final combustible fuel in the trend is hydrogen.

Figure 13-1 shows that fossil fuel production continues to dominate energy production. A reasonable next step in decarbonization would be a transition to natural gas. A greater reliance on natural gas rather than coal or oil would reduce the emission of greenhouse gases into the atmosphere.

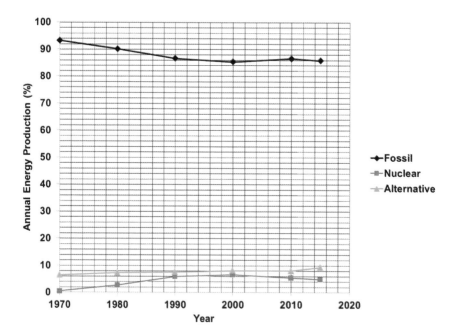

Figure 13-1.   Percent Contribution of Different Energy Types to World Historical Energy Production from 1970 to 2015. [US IEO, 2018]

The trend toward decarbonization depends on technology, infrastructure, and government policy. Fossil fuels (coal, oil, and natural gas) have dominated the energy market in the U.S. for over a century. We know, however, that this reliance cannot be sustained.

The European Union, which has relatively few fossil fuel resources when compared with other regions of the world, is seeking to transition to a renewable energy infrastructure by 2050. Plans call for the EU infrastructure to rely on wind and solar power with a sophisticated electricity transmission grid.

The discovery of new sources of oil and gas give us more time to make a transition to a sustainable energy mix if we assume that catastrophic climate change is not imminent. Our future energy mix depends on choices we make, which depends, in turn, on energy policy. Several criteria need to be considered when establishing energy policy. We need to consider the global capacity of the energy mix, its cost, safety, reliability, and effect on the environment. We need to know that the energy mix can meet our needs (capacity) and be available when it is needed (reliability). The energy mix should have a negligible or positive effect on the environment, and it should be safe for those who work with it and live around it. When we consider cost, we need to consider both tangible and intangible costs associated with each component of the energy mix.

14. Competing energy visions

The challenge we face was expressed by J-M Chevalier in *The New Energy Crisis*: "The challenge of the [21st] century is to provide enough food, water and energy without further damaging the environment" [Chevalier, 2009, page 2]. The availability of food and water resources is going to be impacted by growing population. We can estimate the demand for energy for a growing population if we specify a desired quality of life.

We can estimate how much energy will be needed in the future if we choose a UN HDI corresponding to per capita energy consumption, assume global population growth trends continue, and use the relationship between per capita energy consumption and quality of life characterized by the UN HDI shown in Figure 3-2. For example, if we choose a UN HDI corresponding to 200,000 megajoules energy consumption per person and assume the global population is 8 billion people in 2100, we will need approximately four times as much energy as we are consuming today. This could be a conservative estimate since the assumption of 8 billion people is low compared to population growth forecasts published by the United Nations [UN DESA, 2017]. Figure 14-1 shows three global population forecasts. The "Low" and "Medium" forecasts assume that education will help reduce the growth in population.

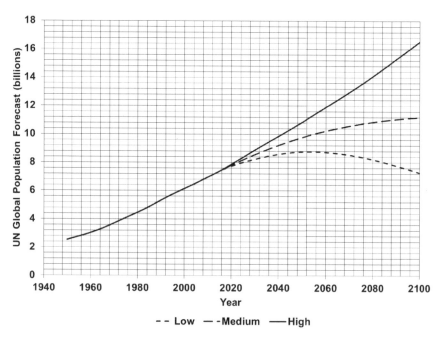

Figure 14-1.   United Nations Global Population Forecast. [UN DESA, 2017]

Energy policy priorities are being discussed within the United States to meet evolving energy demand. For example, former Shell Oil Company President John Hofmeister noted that both major political parties in the United States have called for energy independence, but there has been "no meaningful continuity of political leadership for sufficient sustained periods of time to deliver on the many, many promises of energy independence" [Hofmeister, 2010, page 27]. Hofmeister suggested that effective policies could lead to United States energy independence. He pointed out the need for producing more liquid fuels. He believed that the decline in United States oil production could be reversed by oil and gas development on Federal lands and offshore on the United States outer continental shelf [Hofmeister, 2010, page 29]. This development did not occur during the Obama administration which covered the period from 2009 to 2017, but oil production from shale by the private sector in jurisdictions that allowed production did lead to an increase in United States oil production until the oil price collapse in 2015.

Hofmeister also argued for increases in energy use efficiency and alternative liquid fuels that were not corn-based. He believed that a new agency similar to the United States Federal Reserve Bank would be needed to regulate the energy industry. The new agency, which he called the Federal Energy Resources Board, would regulate four areas: energy supply, technology choices, environmental protection, and infrastructure choices [Hofmeister, 2010, page 197]. Hofmeister's big government solution has not happened. Instead the Trump administration, which began in January 2017, has removed many regulations that were effectively blocking United States energy development by the private sector.

Pundits often say that the United States does not have an energy policy. It is our contention that the United States follows two competing policies, and the choice of policy depends on the political ideology of the party in the White House. The Liberal-Progressive policy says that human activity is driving climate change. Advocates of Liberal-Progressive policy see an urgent need to replace fossil fuels with sustainable or renewable energy sources. Their goal is to save the planet. Advocates of Conservative policy point out that it is important to eventually replace fossil fuels, but the need is not urgent. They believe that the economic health of society is at least as important as possible climate effects. Therefore, Conservatives are willing to transition from fossil fuels to low-carbon energy sources at a relatively slow pace.

The Liberal-Progressive policy, on the other hand, is designed to transition from fossil fuels to low-carbon energy sources at a relatively fast pace. Conservatives would consider the Liberal-Progressive energy transition to be too fast. Advocates of Liberal-Progressive policy emphasize renewable energy sources such as wind and solar energy. To be successful, the cost of renewables must be reduced and energy storage technology must be improved. The EU 2050 plan is relying on a broad geographic distribution of resources and a high-tech electricity transmission grid to minimize the effect of intermittent renewable energy sources. In 2010, former United States Vice President Al Gore [Gore, 2010] noted that President Obama "shifted priorities to focus on building the foundation for a new low-carbon economy." Gore's observation has been justified by the way the Obama administration chose to allocate public resources, as we discuss further in Section 27.

Advocates of the Liberal-Progressive policy consider the Conservative policy to be too slow. Conservatives are willing to extend the country's reliance on fossil fuels and say they would like to develop other resources. They tend to view renewable energy as an emerging technology that is still in the research and development stage. They recognize a need to account for the environmental impact of fossil fuel combustion, but believe there is time.

An alternative approach to the Liberal-Progressive policy and the Conservative policy is the Goldilocks Policy. The idea is to implement a reasonable plan of action that proceeds at just the right pace. We can establish a reasonable transition period by considering historical transitions between a dominant energy source and its replacement. Once a transition period is agreed upon, we can set targets for making the transition. For example, we can make the transition by replacing energy from carbon-based fuels with energy from low-carbon sources by 2 percent per year until we are no longer using carbon-based fuels. In the short-term, we can use natural gas as a transition fuel since natural gas continues the decarbonization trend and reduces greenhouse gas emissions. The Goldilocks Policy replaces uncertain public policy with predictable public policy. It is discussed further in Section 16.

15. How can we transition to a sustainable energy mix?

The conflict between Liberal-Progressive policy and the Conservative policy suggests that well-meaning and well-qualified people can disagree on the impact of human activity on the environment, and how to proceed. Are we prepared to pay the cost of transforming the global economy based on the belief that human activity is changing the climate in adverse ways? What should we do? How long will the transition take? We consider what an energy transition could look like according to three energy experts before presenting the policy proposed here, the Goldilocks policy.

15.1 *John Hofmeister and regulation*

Former Shell Oil Company president John Hofmeister argued for regulation of gaseous waste products on the national level. He was aware that pace was an issue. He observed that if "we rush too fast, it could cost too much. If we legislate in a way that companies and polluters have time to manage their investment future to reduce emissions and do not overly penalize emissions from the start but do so progressively over time, the costs, will be less, the resistance will be lower, and the speed of lowering emissions will be that much faster." [Hofmeister, 2010, page 73]

Hofmeister believed that regulation should apply to four areas: energy supply, technology choices, environmental protection, and infrastructure choices. He suggested that an independent regulatory agency analogous to the United States Federal Reserve Bank should be created. He called the agency the Federal Energy Resources Board (FERB). The FERB would have the power to "determine the amounts and relative percentages of the future supply sources" for the United States [Hofmeister, 2010, page 197]. Technology would have two roles [Hofmeister, 2010, page 199]: "unleash undiscovered or undeveloped new energy supplies and enable far more efficient and clean consumption of energy."

According to Hofmeister the FERB would replace the Environmental Protection Agency as the agency responsible for regulating the impact of energy on the environment in an attempt to minimize political influence. The FERB would have regulatory oversight of "technologies associated with environmental protection; the costs of introducing, monitoring and enforcing such protection; and the impact of such protection on the availability and affordability of supply." [Hofmeister, 2010, page 203]

The fourth area regulated by the FERB would be renovation and updating of infrastructure. The existing energy infrastructure is old and designed for fossil fuels. Hofmeister argued that future infrastructure decisions that impact interstate, regional, or national energy supplies be made by an entity that is authorized to function on the national level, such as the FERB [Hofmeister, 2010, page 204].

15.2 *Vaclav Smil and the transition from fossil fuels to renewables*

Energy scholar Vaclav Smil expressed concern about anthropogenic climate change and his belief that we need to reduce our reliance on fossil fuels [Voosen, 2018]. Smil based his belief on a review of climate data and wrote that climate forecasting models are not accurate. In his 2003 book *Energy at the Crossroads*, Smil provided several examples of failed forecasts and concluded that "long-range forecasters of energy affairs have missed every important shift of the past two generations." [Smil, 2003, page 176] Smil believed that two opposing expectations could explain the inaccurate forecasts. The first expectation is "the spell cast by unfolding trends and by the mood of the moment, the widespread belief that the patterns of the recent past and the aspirations of the day will be dominant, not just in the medium term but even over a long period of time." [Smil, 2003, page 177] The second expectation "is the infatuation with novelties and with seemingly simple, magical solutions through technical fixes" [Smil, 2003, page 177].

Smil has spent decades studying how humanity generates and deploys energy. He said that humanity has passed through three major energy transitions and is now passing through a fourth. The first transition occurred when humanity learned to use fire, that is, the combustion of plants. The second transition occurred when humanity learned to farm. Smil viewed farming as the conversion of solar energy into food. Farming enabled a fraction of humanity to provide food while the rest could pursue other endeavors. The growth of human settlements and the domestication of animals supplied energy in the form of muscle power. The third transition occurred when humanity began to industrialize using fossil fuels. The fourth, and current, transition is to replace energy sources that emit greenhouse gases with energy sources that rely on energy from sunlight, such as wind and solar energy.

The trend toward decarbonization can be expressed in terms of energy density or power density. Smil often writes in terms of power density, which he defined as the ratio of power to land area with International System (SI) metric unit W/m^2 [Smil, 2015, Chapter 2]. Another term for average power per unit area is energy intensity with the same SI unit W/m^2. By contrast, energy density is energy per unit volume with SI unit J/m^3. Energy density can be expressed as specific energy, or energy per unit mass, with SI unit J/kg. Mass density in kg/m^3 is needed to convert from energy density to specific energy.

Historically, humanity has increased the specific energy of its primary energy source as it transitioned from wood, to coal, to oil. Today, humanity is transitioning from high specific energy fossil fuel sources to renewable energy sources. The transition to renewable energy sources increases the footprint needed to provide the needed energy and can have a significant negative impact on the environment. Smil views nuclear fission energy as too risky and costly to be the principal energy source.

Smil observed that energy transitions are slow, difficult and hard to predict. In addition, he wrote in 2003 that humanity needs to accept limits to energy consumption: "the historical evidence is clear: higher efficiency of energy conversions leads eventually to higher, rather than lower, energy use, and eventually we will have to accept some limits on the global consumption of fuels and electricity." [Smil, 2003, page 317] In addition, Smil noted that crises resulting from excessive consumption may not be a harbinger of the collapse of civilization because humans have evolved "to cope with change" [Smil, 2003, page 372]. Smil cautioned in 2015 that "New energy arrangements are both inevitable and desirable, but without any doubt, if they are to be based on large-scale conversions of renewable energy sources, then the societies dominated by megacities and concentrated industrial production will require a profound spatial restructuring of the existing energy system, a process with many major environmental and socioeconomic consequences." [Smil, 2015, page 255]

The current transition from fossil fuels to renewable energy sources has been slowed by the need for technological improvements, such as higher-capacity batteries and more efficient solar photovoltaic (PV) cells. Smil said that humanity could take several decades to transition from fossil fuels to renewables. He wrote in 2003 that the "transition from societies energized overwhelmingly by fossil fuels to a global system based predominantly on conversions of renewable energies will take most of the twenty-first century." [Smil, 2003, page 363] The period of transition could be considerably shortened if a technological breakthrough occurs, such as the development of cheap energy storage.

15.3 *Daniel Yergin and the quest for change*

Daniel Yergin is an economic historian, energy expert, and Pulitzer Prize winning author of *The Prize: The Epic Quest for Oil, Money and Power* [Yergin, 1992]. Geopolitical events in the 1970s precipitated the first oil crises and led to an explicit recognition that there was a need to replace fossil fuels with alternative sources. Yergin saw "a new quest for energy security, a rising environmental consciousness, the specter of permanent 'shortage,' and the assumption that prices would remain permanently high, damaging economic growth." [Yergin, 2013] According to Yergin, there is a renewed sense of urgency today because of concern about climate change and "worry that the current energy mix will not prove adequate to meet the rapidly growing energy needs of emerging market nations." [Yergin, 2013]

Yergin agrees with Vaclav Smil's observation that energy transitions can take decades. To illustrate this observation, Yergin pointed out that the first mechanical steam engine was developed by Thomas Newcomen in 1712, improvements to the steam engine were patented by James Watt in 1775, and Sadi Carnot helped encourage the adoption of the steam engine by publishing a paper explaining how the steam engine worked in 1824 [Yergin, 2015]. The first commercial coal powered steam locomotive appeared in the United States in 1830, more than a century after Newcomen's work.

As another example, Yergin said that solar energy began in 1905 when Albert Einstein published a paper on the photoelectric effect, which was an important aspect of the theoretical basis of solar PV technology. The first silicon based solar cell was developed in the 1950s, half a century later, to provide electricity for satellites in what was emerging as the space race between the Soviet Union and the United States.

Solarex, the first company to produce commercial solar cells, was founded in 1973 before there was an established market for solar panels. A few customers began to appear that were interested in providing electricity in remote locations, such as offshore buoys and offshore platforms. There was not enough business to achieve a sustainable economic scale, and the solar industry was almost dead by the 1990s. This changed in the first decade of the 21st century when Germany decided to develop a low-carbon economy.

Germany "launched its Energiewende (energy transition), which provided rich subsidies for renewable electricity." [Yergin, 2015, page 7] Low-cost manufacturing facilities for solar cells were established in China. The overcapacity of Chinese factories and the falling prices of silicon, a key component of modern solar cells, pushed down costs and the increased economic competitiveness of solar cells. The combination of declining costs, expanding capacity, government subsidies and regulations has established a market for solar cells in the present decade.

These examples demonstrate that it can take a century to go from innovation to commercial production of an energy technology. The length of time in the examples begins with discovery. Many renewable energy technologies today are relatively mature, and the time for achieving transition to new energy sources will be shortened by the existing level of maturity of the technology. We show in Section 16 that the duration of the transition period for known energy sources has historically ranged from 60 to 70 years.

The energy mix of the future will depend in large part on price and value delivered. Price will depend on such factors as competitive marketplace, price on carbon, and government policy. As examples, Yergin wrote that government policies have been designed to spur demand for electricity from renewable energy sources and support the commercialization of electric vehicles [Yergin, 2013].

Historically, shifts in energy from one dominant source to another can take decades. In addition, the older energy source will not necessarily disappear, but co-exist with the newer source. For example, the global energy mix still includes coal and wood even though oil is the primary energy source for transportation.

16. Goldilocks policy for energy transition

Concerns about the environmental impact of combustible fuels and security of the energy supply are encouraging a movement away from fossil fuels. Political instability in countries that export oil and gas, the finite size of known oil and gas supplies, increases in fossil fuel prices and decreases in renewable energy costs, such as wind energy costs, are motivating the adoption of renewable energy. On the other hand, the development of technology that makes unconventional sources of fossil fuels economically competitive is encouraging continued use of fossil fuels, especially as the natural gas infrastructure is improved. These conflicting factors have an impact on the rate of transition from fossil fuels to a sustainable energy mix. A key decision facing society is to determine the rate of transition. In Section 15 we reviewed the opinions of some experts that the transition could last the rest of the 21st century. Here we use the historical basis to determine a reasonable transition period from one known energy source to another. We then present the Goldilocks Policy.

16.1 *Historical basis for duration of energy transitions*

The United States is a developed country with a history of energy transitions over the past few centuries. Figure 16-1 shows United States energy consumption by source beginning in 1650. Wood was the principal energy source when the U.S. was founded in the 18th century. Coal began to take over the energy market in the first half of the 19th century and

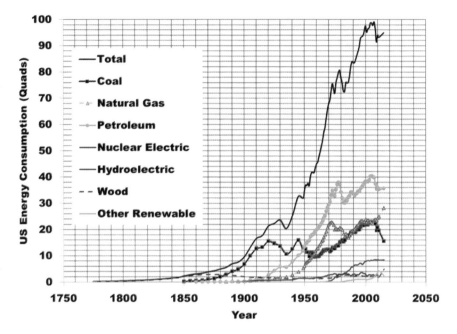

Figure 16-1. Can History Tell Us How Long an Energy Transition Period Will Last? [http://www.eia.gov/totalenergy/data/annual/perspectives.cfm and Annual Energy Review 2001, Appendix F, Tables F1a and F1b; Fanchi, 2013]

peaked in the early 20th century. Oil began to compete with coal during the latter half of the 19th century and became the largest component of the energy mix by the middle of the 20th century.

Historical transition periods can be highlighted by presenting energy consumption by source as a percentage of total United States energy consumption. Figure 16-2 shows the transition periods from wood to coal and from coal to oil. The United States was dependent on wood until around 1850 when the transition to coal began. The United States depended on coal until 1900–1920. The transition to oil was stimulated by the First World War which saw the development and use of vehicles powered by the internal combustion engine. Today, oil and natural gas account for over 80% of energy consumption. Historically, the coal and oil transition periods lasted between 60 and 70 years. The historical energy

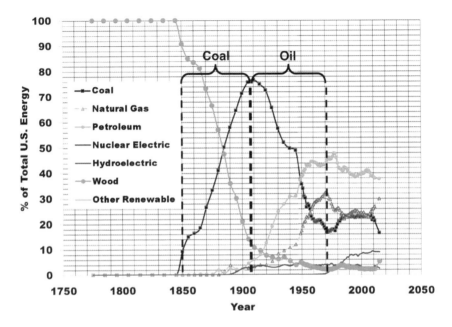

Figure 16-2. Coal and Oil Transition Periods Based on U.S. Energy Consumption by Source, 1650–2010. [Fanchi, 2013; and http://www.eia.gov/totalenergy/data/annual/perspectives.cfm and Annual Energy Review 2001, Appendix F, Tables F1a and F1b]

transition periods in the United States give us an idea of how long it has historically taken a mature society to make the transition from one energy source to another energy source.

16.2 *Temperature change forecast*

There is a concern that catastrophic climate change may be imminent. This raises the question: do we have time to make a transition to sustainable energy sources? The COP21 Paris Climate Agreement was an agreement to reduce emissions with a long-term goal of keeping the increase in global average temperature to well below 2°C (3.6°F) above pre-industrial levels. [COP21 Agreement, 2015] How can we determine when temperature change will exceed 2°C?

Global climate models can be used to estimate the timing of temperature change. Here we use an alternative, empirically based method that recognizes an apparent correlation between global population and temperature change. It is shown in Figure 16-3 as the regression. The other three cases are projections from the global climate dashboard of the United States National Oceanic and Atmospheric Administration [US NOAA Temperature, 2018]. The projections are averages of global climate models for low, moderate and high growth. The regression case is comparable to the low growth case.

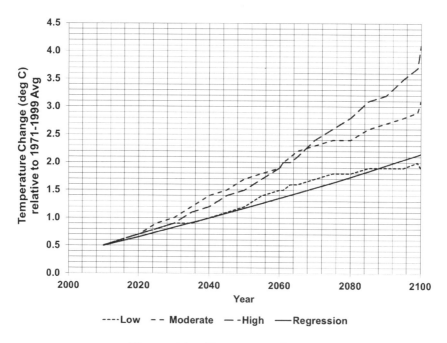

Figure 16-3. Temperature Change.

The temperature change relative to the average of temperatures from 1971 to 1999 reaches 2°C sometime between 2060 and 2100 in Figure 16-3. The baseline temperature calculated for the period ranging from 1971 to 1999 is a relatively higher temperature than an average pre-industrial temperature. This implies that the baseline temperature is too high to satisfy the COP21 Paris Climate Agreement, but it gives a more modern baseline for comparison. If we use the 2°C temperature change relative to the 1971–1999 average, we see that the 2°C temperature change could occur in as few as 40 years or after the turn of the century in 2100.

16.3 *The Goldilocks policy*

Fanchi and Fanchi [2015] introduced another proposal that makes it possible to gradually transition to a sustainable energy mix. The Goldilocks policy may be viewed as a Grand Energy Bargain between the Liberal-Progressive policy and the Conservative policy. Concerns about Russian adventurism in Ukraine and European dependence on Russian fossil fuels motivated left-wing columnist Thomas L. Friedman to suggest the adoption of a Grand Energy Bargain advocated by Hal Harvey of Energy Innovation. The Friedman-Harvey Grand Energy Bargain [Friedman, 2014] would advance economic growth, provide for national security, and recognize environmental concerns. Key ingredients of United States energy strategy in the Friedman-Harvey Grand Energy Bargain include:

1. Simultaneously optimize energy affordability, reliability and environmental compatibility
2. Use modern technology to provide affordable, reliable and clean energy
3. Use government to ensure that natural gas resources are used to usher in a secure, clean-energy future

The Friedman-Harvey Grand Energy Bargain would include four steps:

1. Adopt national rules for extracting natural gas
2. Set a national clean energy standard for electricity, e.g. utilities must implement zero-carbon sources (wind, solar, nuclear fission) at a predictable rate, such as 2% per year

3. Accelerate energy efficiency and clean energy technology by increasing research and development
4. Replace payroll and corporate taxes with a revenue-neutral carbon tax

The details of the Friedman-Harvey Grand Energy Bargain would have to be negotiated. For example, Hofmeister's Federal Energy Resources Board could be used to implement some of the steps of the Friedman-Harvey Grand Energy Bargain, but Hofmeister expressed support for a cap-and-trade system rather than a carbon tax. In addition, limited government advocates might object to some of the details, such as a carbon tax, a federal agency such as the FERB with extensive regulatory power, or adoption of a cap-and-trade system.

The Goldilocks Policy calls for the reduction or replacement of fossil fuel consumption by 2% per year. It is possible to forecast energy consumption based on the Goldilocks Policy given a few assumptions. For example, suppose we assume that there is linear growth in energy consumption based on a linear fit of consumption in this century, and that there is no change in the consumption of nuclear fission energy. The Goldilocks policy is satisfied by increasing the consumption of alternative energy by 2% per year to match the reduction in fossil fuel consumption. If we begin implementing these assumptions in 2020, we obtain the energy consumption forecast shown in Figure 16-4. According to this forecast, fossil fuel consumption would end by 2080.

A wild card in this discussion is the development of nuclear fusion energy. Several countries are sponsoring the International Thermonuclear Experimental Reactor (ITER) being built in Southern France. The nuclear fusion reactor is a tokamak design that is expected to produce 500 MW fusion power from 50 MW input power. People have been working to develop commercial fusion power for decades, so no one can say with certainty when nuclear fusion power technology will be ready for peaceful commercial use. Other nuclear fusion concepts are being studied. For example, Lockheed Martin is developing a compact fusion reactor which

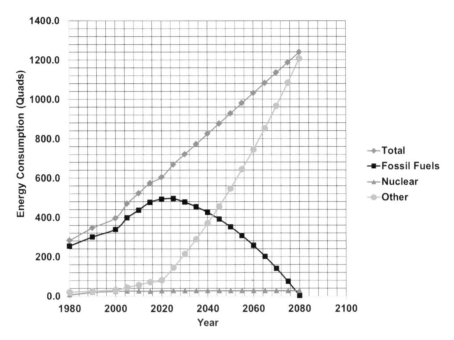

Figure 16-4.   Forecast of Energy Consumption Based on the Goldilocks Policy.

is a nuclear fusion reactor on a smaller scale than ITER, and German researchers at the Max Planck Institute are developing a nuclear fusion reactor known as the Wendelstein 7-X stellarator. The successful development and commercialization of nuclear fusion power would provide a virtually inexhaustible source of energy that is environmentally benign, but we cannot count on nuclear fusion.

The Goldilocks policy is a roadmap to a sustainable energy future. It would require discipline and patience, but it would enable us to make an orderly transition from non-renewable, combustible, carbon-based energy sources to a sustainable economy. There are obstacles, however, that should be considered.

Part IV

Obstacles to Adopting the Goldilocks Policy

17. Conventional political obstacles

The Goldilocks Policy is a policy that will require decades to implement. There are many obstacles that can block implementation of the Goldilocks Policy, or change the policy after it has been adopted. We must consider the role of historical trends to understand the obstacles and their potential impact on energy policy. The trends include environmentalism, socialism, and globalism. These trends have many intersections that support the idea that anthropogenic climate change is, in part, being used as a tool to justify the adoption of a supranational authority controlled by a ruling class that seeks to impose an urgent transition from fossil fuels to renewable energy. We begin our discussion by considering conventional political obstacles.

17.1 *Typical forms of government*

Competing energy visions have influenced United States energy policy. The implementation of an energy policy depends on the type of government in a society. In this section, we review different forms of government before discussing the role of government in the energy sector.

The terms "left" and "right" arose in politics from the seating arrangement of political parties in the European Parliament. If we stand at the podium and look out over the audience, we would see members of Parliament with liberal, socialist, green, or communist views on the left side of the podium, and conservatives, Eurosceptics, and anarchists on the right side of the podium. Modern liberalism tends to seek social justice by supporting the use of government taxation and social programs to redistribute wealth. Socialism and Communism seek greater control of the means of production by increasing central planning and regulation. Green party members support government regulation of activities that can affect the environment. By contrast, modern conservatives support capitalism and limited government regulation. Eurosceptics are critical of the European Union governance system and believe that EU governance threatens national sovereignty. Anarchists support minimal or no government control with political power vested in the individual.

Obstacles to Adopting the Goldilocks Policy ▎ 85

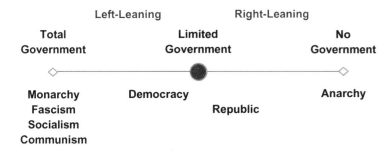

Figure 17-1. Spectrum of Governmental Control.

Figure 17-1 illustrates the range of political beliefs in terms of government control. The "far left" represents total governmental control while the "far right" represents no governmental control. A ruling class tends to accumulate political power as you move to the left while political power tends to be vested in the individual as you move to the right.

The role of government depends on the form and size of government. Citizens exercise political power directly in a democracy. Representatives are selected to exercise political power in a republic. The United States was founded as a constitutional republic, that is, a republic with the powers of government enumerated in a written constitution. In a monarchy, a hereditary chief of state serves for life and exercises power ranging from nominal to absolute. Political authority is centralized under a dictator in fascism. The dictator can impose stringent government controls. Political power is centralized in a ruling class in socialism, communism, and oligarchic political systems discussed below.

The size of government can be quantified by expressing government spending in terms of the size of the economy. For example, government spending by the United States Federal government ranged from approximately 18% of GDP to 25% of GDP during the first decade of the 21st century. As a rule of thumb, governments to the left in Figure 17-1 tend to spend a larger percentage of national GDP than those to the right. Government expenditures during times of war can alter this observation. The amount of money spent directly by the government is not the same as the amount of resources controlled by the government through policies,

17.2 Government and energy

Governments influence the energy sector. The political right in the United States has argued that the government should not play much, if any, role in the development of clean energy. Their argument is that the free market will determine which technologies are best used and when transitions should occur. As fossil fuel resources are depleted, they argue, the price will increase and the market will be inclined to seek alternatives such as wind and solar. The counter-argument is that without government attaching a cost to pollution and greenhouse gas emissions, the market would not account for them and the environment would suffer.

The political left, on the other hand, believes that government should tax and regulate industries that pollute and emit carbon. This factors environmental costs into the cost of the resource. In addition, the left believes that the government can play a role in supporting clean technologies in their infancy when they are too expensive and risky for private enterprises to research and develop. This is where direct government investment (Solyndra) and indirect subsidies (Spanish solar power) come in. The counter-argument to this approach is that government intervention in the free market will distort prices and will tend to artificially inflate or suppress prices in a way that is unsustainable. Government has limited resources, and those resources could potentially be better spent elsewhere. Finally, government involvement creates the potential for fraud, corruption and political favoritism.

17.3 Mass media

Mass media is a powerful tool for political parties to use to shape public opinion on politically volatile issues. Mass media and social media have significantly impacted world events in the 21st century. Some of the events include Presidential Elections in the United States and the 2010–12 Arab Spring. Energy is a topic that receives significant media coverage in the United States. Energy issues such as climate change and fracking often lead news programs, are the subject of numerous documentaries, and find their way into fictional films and TV programs. From hard news programs to satirical news to animated comedy shows, Americans are frequently presented with perspectives on energy topics that are carefully crafted to further political agendas. In a society where the population's attention span is undeniably diminishing and the public is frequently getting their news from non-news sources, crafting eye-catching headlines and integrating political perspectives into non-political programming has become a critical aspect of the war to win public opinion.

Social networks are a tool for influencing public opinion. For example, Donald Trump used Twitter messages limited to 140 characters to express his views in the 2016 United States Presidential campaign. He continues to use Twitter during his Presidency to present topics for discussion in daily news cycles.

18. Role of geopolitics

Geopolitics is concerned with relationships between nations. Here we review several perspectives on the way nations interact with one another. This review is designed to help us better understand historical and current events, and provides a foundation for understanding the socio-political issues that affect energy demand.

18.1 *Clash of civilizations*

The world has been undergoing a socio-political transition that began with the end of the Cold War and continues today. S.P. Huntington [1996] argued that a paradigm shift was occurring in the geopolitical arena.

The Cold War between the Soviet Union and the Western alliance led by the United States established a framework that allowed people to better understand the relationships between nations following the end of World War II in 1945. When the Cold War ended with the fall of the Berlin Wall and the break-up of the Soviet Union in the late 1980's and early 1990's, it signaled the end of one paradigm and the need for a new paradigm. Several geopolitical models have been proposed. Huntington considered four possible paradigms for understanding the transition (Table 18-1).

Table 18-1. Huntington's Possible Geopolitical Paradigms.

1	One Unified World
2	Two Worlds (West versus non-West; us versus them)
3	Anarchy (184+ Nation-states)
4	Chaos

The paradigms in Table 18-1 cover a wide range of possible geopolitical models. The One Unified World paradigm asserts that the end of the Cold War signaled the end of major conflicts and the beginning of a period of relative calm and stability. The Two Worlds paradigm views the world in an "us versus them" framework. The world was no longer divided by political ideology (democracy versus communism); it was divided by some other issue. Possible divisive issues include religion and rich versus poor (generally a North-South geographic division). The world could also be split into zones of peace and zones of turmoil. The third paradigm, Anarchy, views the world in terms of the interests of each nation, and considers the relationships between nations to be unconstrained. These three paradigms, One Unified World, Two Worlds, and Anarchy, range from simple (One Unified World) to complex (Anarchy).

The Chaos paradigm says that post-Cold War nations are losing their relevance as new loyalties emerge. In a world where information flows freely and quickly, people are forming allegiances based on shared traditions and value systems. The value systems are notably cultural and, on a more fundamental level, religious. The new allegiances are in many cases a rebirth of historical loyalties. New alliances are forming from the new allegiances and emerging as a small set of civilizations. The emerging civilizations are characterized by ancestry, language, religion, and way of life.

Huntington considered the fourth paradigm, Chaos, to be the most accurate picture of current events and recent trends. He argued that the politics of the modern world can be best understood in terms of a model that considers relationships between the major contemporary civilizations shown in Table 18-2. The existence of a distinct African civilization has been proposed by some scholars, but is not as widely accepted as the civilizations identified in the table.

Table 18-2. Huntington's Major Contemporary Civilizations.

Civilization	Comments
Sinic	China and related cultures in Southeast Asia
Japanese	The distinct civilization that emerged from the Chinese civilization between 100 and 400 A.D.
Hindu	The peoples of the Indian subcontinent that share a Hindu heritage.
Islamic	A civilization that originated in the Arabian peninsula and now includes subcultures in Arabia, Turkey, Persia, and Malaysia.
Western	A civilization centered around the northern Atlantic that has a European heritage and includes peoples in Europe, North America, Australia, and New Zealand.
Orthodox	A civilization centered in Russia and distinguished from Western Civilization by its cultural heritage, including limited exposure to Western experiences (such as the Renaissance, the Reformation, and the Enlightenment).
Latin America	Peoples with a European and Roman Catholic heritage who have lived in authoritarian cultures in Mexico, Central America and South America.

Each major civilization has at least one core state [Huntington, 1996, Chapter 7]. France and Germany are core states in the European Union, which is viewed as part of Western Civilization. The United States is also a core state in Western Civilization. Russia and China are core states, perhaps the only core states, in Orthodox Civilization and Sinic Civilization, respectively. Core states are sources of order within their civilizations. Stable relations between core states can help provide order between civilizations.

The growth of multiculturalism in some states has established communities within those states that may not share the values and allegiances of the host state. A multicultural state in this context is a member state of one civilization that contains at least one relatively large group of people that is loyal to many of the values of a different civilization. For example, Spain is a member state of Western Civilization with a sizable Islamic population. Other European Union member states have become more multicultural due to the influx of foreigners during the first two decades of the 21st century. Many of the foreigners are Moslems that were displaced by war in the Middle East. After the collapse of the Soviet Union, some multicultural states (e.g. Yugoslavia and Czechoslovakia) that were once bound by strong central governments separated into smaller states with more homogeneous values.

Within the context of the multi-civilization geopolitical model, the two World Wars in the 20th century began as civil wars in Western Civilization and engulfed other civilizations as the hostilities expanded. Those wars demonstrate that civilizations are not monolithic: states within civilizations may compete with each other. Indeed, the growth of multiculturalism and large migrant populations in some states are making it possible for states within a civilization to change their cultural identity as cultures within member states compete for dominance. Today the presence of large Islamic communities is more widespread in what is now a relatively secular Western civilization that has practiced an open borders policy. The change in demographics can lead to a change in cultural identity and a change in allegiance to a civilization.

The Cold War and the oil crises in the latter half of the 20th century were conflicts between civilizations. Western Civilization has been the most powerful civilization for centuries, where power in this context refers to the ability to control and influence someone else's behavior. The trend in global politics is a decline in the political power of Western Civilization as other civilizations develop technologically and economically. Energy is a key factor in this model of global politics. This can be seen by analyzing the energy dependence and relative military strength of core states. For example, consider the relationship between Western and Islamic Civilizations.

Western Civilization is an importer of oil and many states in Islamic Civilization are oil exporters. The result is the transfer of wealth from oil importing states of Western Civilization to oil exporting states in Islamic Civilization. By contrast, the United States, a leading core state in Western Civilization, is the leading military power in the world with a large arsenal of nuclear weapons. Most core states in Islamic Civilization are relatively weak militarily and do not have nuclear weapons. The wealth being acquired by Islamic Civilization is being used to alter the balance of military power between Western Civilization and Islamic Civilization. Iran, a core state in Islamic Civilization, is using its oil wealth to improve its arsenal of conventional weapons and acquire nuclear technology from core states in other civilizations.

Ideological differences between civilizations can lead to a struggle for global influence between core states. The battlefields in this struggle can range from economic to ideological to military. The outcome of this struggle is influenced by access to energy.

Energy importing states in one civilization rely on access to energy sources from energy exporting states in other civilizations. If the relationship between energy trading states is hostile, energy becomes a weapon in the struggle between civilizations. For example, the growth of non-Western civilizations, such as Sinic and Hindu Civilizations, has increased demand for a finite volume of oil. This increases the price of oil as a globally traded commodity and increases the flow of wealth between oil importing civilizations and oil exporting civilizations. Oil importing nations may try to reduce their need for imported oil by finding energy substitutes or by conservation. The social acceptability of energy conservation varies widely around the world. In some countries, such as Germany, energy conservationists and environmentalists are a political force (the Green Party). In other countries, such as the United States, people may espouse conservation measures but may be unwilling to participate in or pay for energy conservation practices, such as recycling, carpooling, or driving energy-efficient vehicles.

Energy production depends on the ability of energy producers to have access to natural resources. Access depends on the nature of relationships between civilizations with the technology to develop natural resources and civilizations with territorial jurisdiction over the natural resources. Much oil production technology was developed in Western Civilization and gave Western Civilization the energy it needed to become the most powerful civilization in the world. As Western Civilization consumed its supply of oil, it became reliant on other civilizations to provide enough oil to maintain its oil-dependent economies. This dependence in times of stress between civilizations can lead to social turmoil and conflict between states that are members of different civilizations. Impending turmoil is motivating member states to develop alternatives to fossil fuels in an effort to achieve energy independence. The EU Supergrid is a program designed to give the EU energy independence by 2050 by transitioning from fossil fuels to renewable sources of energy (wind, solar, wave, and geothermal). The development of shale oil and gas resources has increased the economic supply of fossil fuels in the United States and reduced the influence of OPEC.

18.2 Clash over resources

Huntington's Clash of Civilizations provides one perspective on modern geopolitical events. Another perspective has been advocated by Michael T. Klare [2004], who argued that modern geopolitics is driven by a struggle for resources. Klare summarized his thesis as follows [Klare, 2004, page xii]: "After examining a number of recent wars in Africa and Asia, I came to a conclusion radically different from Huntington's: that resources, not differences in civilizations or identities, are at the root of most contemporary conflicts." In *Resource Wars* [Klare, 2001], Klare made the case that oil, water, land and minerals were each important enough to be a source of contention. He narrowed the list to petroleum in *Blood and Oil* [Klare, 2004]. Inexpensive and abundant petroleum is essential to modern economies, especially in the United States where energy based on fossil fuels has supported the American lifestyle. In general, Klare concluded that Huntington's cultural differences between civilizations may not be as important in determining geopolitical relationships as much as who controls valuable resources.

18.3 Energy interdependence

Some people have argued that the world can continue to rely on fossil fuels for decades to come. Robin Mills, petroleum economics manager at Emirates National Oil Company in Dubai, said that peak oil supply has not yet been reached [Mills, 2009]. At the time, the main cause for concern about oil supply was that oil prices were too low to encourage exploration.

Oil and gas production technology has improved significantly since the first oil crisis in 1973. For example, seismic resolution has improved our image of the subsurface, multiple seismic surveys allow us to see how fluids move and where commercial deposits of oil and gas may be, completion techniques such as hydraulic fracturing have made resources in low permeability formations producible, and extended reach drilling of multilateral wells reduces the footprint and environmental impact of drilling. These technologies can be economically justified when the price of oil and gas are high enough, but the price needs to be sustained long enough to make a project commercially feasible.

The search for and development of new fossil fuel resources can be risky and expensive. Another source of resources is already well known: fields that have already been produced and abandoned often contain between one third and two thirds of their original resources. New techniques and higher prices may make many of these old resources commercial. Mills concluded that the "best place to look for new oil is in old fields, with fresh ideas" [Mills, 2009, page 17].

Mills noted that the public has a negative perception of the oil industry. Claims that peak oil supply is imminent are increasing political pressure to prematurely shift away from fossil fuels. Furthermore, many young people were choosing career paths outside of the oil and gas industry rather than enter a "sunset industry." The lack of interest in the oil and gas industry by top graduates has been an industry concern for decades. For example, oil prices have historically tended to rise and fall in cycles. During one cycle, low oil prices in the 1980's led corporations to reduce their workforces with layoffs, which helped support the negative perception that oil and gas industry employees were expendable.

Mills suggested that the best way to achieve global energy security is for countries to recognize that a large energy-consuming nation would probably be unable to achieve energy independence at an acceptable cost. Instead, Mills argued that "energy security can only be achieved, or at least improved, by a balance between needs of exporters and importers, and a web of mutual interdependency" [Mills, 2009, page 17].

Another issue to consider is the argument that combustion of fossil fuels causes so much environmental damage that it must be stopped. While methods of decreasing the environmental impact of burning fossil fuels are improving, such as carbon capture and storage, some authors consider these methods inadequate for handling the large volume of greenhouse gases generated by societies that rely on fossil fuels. People who seek to increase the cost of fossil fuels by imposing a tax on its use argue that the cost of using fossil fuels does not include the cost to mitigate environmental damage. They believe the cost of fossil fuel combustion is understated and must be increased. An increase in the cost of fossil fuels would make renewable energy options more attractive economically.

19. The political roots of socialist environmentalism

Conventional political systems and their geopolitical interaction were discussed in Sections 17 and 18. We showed that governments and geopolitics can impact energy policy and be obstacles to implementing the Goldilocks Policy for energy transition. In this section we consider another trend that can impact energy policy: the possibility that socialism and environmentalism are being used as political weapons.

19.1 *Is modern environmentalism an attempt to impose socialism?*

Brian Sussman made the case that modern environmentalism is attempting to impose socialism on a global scale [Sussman, 2012]. He traced the beginning of the movement back to Karl Marx (1818–1883), a proponent of organized collectivism, which is also known as socialism or communism.

Marx was influenced by the philosophy of Georg Hegel (1770–1831) while Marx was a student at the University of Berlin. Hegel believed that a new world-view based on scientific reason and fact was needed to replace religions that relied on beliefs and faith, notably Judaism and Christianity.

Judeo-Christian beliefs make a statement about the value of property and the role of humans in history [Smith, 1991, Chapter VII]. The first line of Genesis — the first book of the Old Testament — expresses the belief that God created the material world: "In the beginning God created heaven and earth." [Torah, 1992, Genesis 1:1; see also RSV *Bible*, 1971, Genesis 1:1] Furthermore, Genesis says that "God saw all that He had made, and found it very good." [Torah, 1992, Genesis 1:31; see also RSV *Bible*, 1971, Genesis 1:31] The notions that the existing material world was created by God and is considered good implicitly attributes worth to material existence. Furthermore, the statement in Genesis that humans should "fill the earth and master it; and have rule over the fish of the sea, the birds of the sky, and all the living things that creep on earth" [Torah, 1992, Genesis 1:28; see also RSV *Bible*, 1971, Genesis 1:28] implies that property ownership is sanctioned by God.

The Genesis version of creation asserts that the arrival of man in the universe was planned. Judaism teaches that the events of history have a purpose decreed by God [Smith, 1991, Chapter VII]. Attributing meaning to history gives significance to the activities, institutions, and social order of humanity because they can influence the course of history. Anything capable of affecting history can aid or deter the fulfillment of God's purpose.

Property and the role of humanity in history were topics of concern to Marx and his colleague Frederick Engels (1820–1895). The two men had similar views on dialectics, materialism, and religion. The word dialectic was used by Plato (ca. 428–348 B.C.) in the Socratic dialogues. The dialectical process introduced by Hegel refers to discussion between people with different points of view who are seeking truth using reasoning. Dialectic is not the same as debate because debate can use any method to persuade while dialectic is limited to reasoned argument. Hegel used the dialectical process to seek a synthesis of different points of view represented by a thesis and its antithesis. In the 1920's, Marx and Engels combined concepts from materialism and dialects into dialectical materialism.

19.2 *Dialectical materialism*

Marx wrote his doctoral dissertation in philosophy in 1841 [Marx, 1841]. The dissertation was entitled "The Difference between the Democritean and Epicurean Philosophy of Nature." Classical Greek philosopher Epicurus believed that matter was composed of invisible, indivisible bits of matter called atoms. His theory of matter was known as materialism in the 1800s.

Epicurus taught that the physical world consisted of matter and there was nothing else. Marx and Engels shared this view. In 1886, Engels wrote that "if science can get to know all there is to know about matter, we will then know all there is to know about everything" [Engels, 1886]. This view is shared by many modern scholars. In his 20th century television series Cosmos (1980), astronomer and atheist Carl Sagan agreed [Sagan, 2005]: "The cosmos is all there is, or was, or ever will be."

Dialectical materialism is a combination of the concepts of dialectics and materialism. According to dialectical materialism, everything is composed of matter, and a material system can attain a more stable state by resolving conflict associated with opposing qualities.

Marx and Engels formulated three Laws of Matter: Law of Opposites; Law of Negation; and Law of Transformation. The Law of Opposites asserts that nature contains objects with opposing characteristics. Humans, for example, exhibit opposing qualities like love and hate, courage and cowardice, etc. Physical examples of the Law of Opposites include matter-antimatter, and positive-negative electric charge. Objects with opposing characteristics co-exist when equilibrium is achieved. In the human world, conflict could arise between competing ideas and events, and government is the only institution powerful enough to resolve human conflict.

Taoism adopts a view of human behavior similar to the Law of Opposites within the context of the Yin-Yang depicted in Figure 19-1.

98 | The Goldilocks Policy

Figure 19-1. Yin-Yang.

The Yin and the Yang represent polar opposites. Feminine traits, intuitive knowledge, passive responses, and evil behavior are typically represented by the Yin; whereas Yang depicts their opposites, namely masculine traits, rational knowledge, active responses, and good behavior. The dark dot in the Yang and the light dot in the Yin are symbolic of the Chinese concept that the Yin and the Yang always contain some semblance of their opposite, even when they are at their peaks. The pairs of concepts symbolized by the Yin-Yang are complementary.

The Law of Negation says that species can proliferate by negating themselves, or dying. This allows one generation of a species to make way for a succeeding generation and makes it possible for the species to reproduce in greater numbers. The population explosion today shows that humans are not very successful at self-regulation, which can be interpreted to imply that government may need to intervene to assure sustainability.

The Law of Transformation says that species can change (evolve). The change was viewed as progressive, or towards an improvement in the species. In the process, one part of the species may make a significant evolutionary leap while co-existing with the less developed part of the species. The Law of Transformation can be interpreted to imply that an elite status may be conferred upon a class of humans and leads to the concept of organized collectivism: the idea that humans should have a superior ruling class that is expected to care for the species as a whole.

Sussman says the three Laws of Matter "provide the rationale for today's green agenda" [Sussman, 2012, page 3]. Furthermore, the reliance on humanity and the rejection of the supernatural lead to a system that embraces relativism and rejects absolutism expressed in the form of absolute truth.

19.3 *The Marxist view of property*

Hegel and Marx rejected the founding concepts expressed in the American Declaration of Independence (1776), which declared that "We hold these truths to be self-evident, that all men are created equal, that they are endowed by their Creator with certain unalienable Rights, that among these are Life, Liberty and the pursuit of Happiness." [Founders, 1776] Hegel believed that freedom must be granted by the state: "The state as a completed reality is the ethical whole and the actualization of freedom. It is the absolute purpose of reason that freedom should be actualized." [Hegel, 1820, page 197] Similarly, Marx did not believe that the life of the individual had value except within the collective. Furthermore, Marx believed that "philosophers have only interpreted the world, in various ways; the point, however, is to change it." [Marx, 1845, Thesis XI] The idea that humans had rights endowed by God to life, liberty and the pursuit of happiness were contrary to the beliefs of an atheistic materialist that believed in the collective and wanted to change the world.

By contrast, the American Declaration of Independence was influenced by political philosopher John Locke (1632–1704), who wrote that the "necessity of pursuing happiness [is] the foundation of liberty." [Locke, 1690, Chapter XXI, paragraph 52] Furthermore, Locke provided a justification for the private ownership of property by endeavoring "to show how men might come to have a property in several parts of that which God gave to mankind in common" [Locke, 1680, paragraph 25,

Chapter 5 entitled "Of Property", 2nd treatise]. Locke argued that "every man has a 'property' in his own 'person.' ... The 'labour' of his body and the 'work' of his hands are properly his. Whatsoever, then, he removes out of the state that Nature hath provided and left it in, he hath mixed his labour with it, and joined to it something that is his own, and thereby makes it his property." [Locke, 1680, paragraph 27, Chapter 5, 2nd treatise]

Property may include tangible items such as land, merchandise, or money. It can also include intangibles such as a person's opinions and the right to express them. In a 1792 Essay entitled "Property", American founder James Madison clarified the meaning of property by stating that property "embraces everything to which a man may attach a value and have a right; and which leaves to everyone else like advantage." [Madison, 1792]

Marx believed that capitalism led to class struggle between the bourgeoisie and the proletariat. The bourgeoisie are the owners and controllers of the means of production, while the proletariat consists of laborers. Marx and Engels wrote: "The distinguishing feature of Communism is not the abolition of property generally, but the abolition of bourgeois property. But modern bourgeois private property is the final and most complete expression of the system of producing and appropriating products that is based on class antagonisms, on the exploitation of the many by the few.

"In this sense, the theory of the Communists may be summed up in the single sentence: Abolition of private property." [Marx and Engels, 1848, Section II, page 223]

19.4 *The Marxist view of natural resources*

According to Sussman [Sussman, 2012, page 8], the first example of a scientist attacking capitalism based on environmental impact was German chemist Justus Liebig (1803–1873), a pioneer in organic chemistry. Liebig observed that the use of guano (bird droppings) as a fertilizer by capitalist agriculturalists was leading to unexpected and undesirable environmental effects. He said that the "barren soil on the coast of Peru is rendered fertile by means of a manure called guano, which is collected from several islands on the South Sea." [Liebig, 1840, page 74] In a footnote on the same page, Liebig explained that guano "forms a stratum several feet in thickness upon the surface of these islands, [and] consists of the putrid excrements of innumerable sea-fowl that remain on them during the breeding season." The nutrient value of phosphate-rich guano made it an important agricultural additive in many parts of the world, including Europe and the Americas.

The extensive use of guano became a concern for Liebig, who believed that a society that took nutrients out of the soil and did not restore them was engaged in a robbery system. He supported the principle that all nutrients taken from the soil had to be restored, and believed that "the commercial farming system of contemporary Europe violated this principle." [Marold, 2002, page 73] Liebig illustrated the principle by observing that Great Britain, a leader in European agriculture, "deprives all countries of the conditions of their fertility. It has raked up the battle-fields of Leipsic, Waterloo and the Crimea; it has consumed the bones of many generations accumulated in the catacombs of Sicily; and now annually destroys the food for a future generation of three millions and a half of people. Like a vampire it hangs on the breast of Europe, and even the world, sucking its lifeblood without any real necessity or permanent gain for itself." [Märald, 2002, page 74]

Liebig's work influenced Marx's views on ground rent. Ground rent is an agreement between a landlord and a tenant which authorizes the tenant to pay the landlord for the right to use a plot of land. The tenant may own property on the land. In a letter to Engles, Marx said [Marx, 1866]: "I had to plough through the new agricultural chemistry in Germany, in particular Liebig and Schönbein, which is more important for this matter [ground rent] than all the economists put together." According to Marx, one of Liebig's immortal merits was to have "developed from the point of view of natural science, the negative, i.e., the destructive side of modern agriculture." [Marx, 1887, page 357, note 246]

Marx said that "all progress in capitalistic agriculture is a progress in the art, not only of robbing the laborer, but of robbing the soil; all progress in increasing the fertility of the soil for a given time, is a progress towards ruining the lasting sources of that fertility." [Marx, 1887, page 330] The "moral of history," concluded Marx, "is that the capitalist system works against a rational agriculture, or that a rational agriculture is incompatible with the capitalist system." [Marx, 1894, page 83]

19.5 The Marxist view of environmentalism goes national

Marx believed that natural resources did not belong to anyone and should only be used for the common good. His views on the environment were passed on to modern environmentalists through a succession of proponents. Foster traced the development of a socialist view of natural history through Britain where dialectical materialism, Marx, Darwin, and science were linked [Foster, 2002, page 81].

Edwin Ray Lankester (1847–1929) was a British zoologist, Darwinist, and protégé of Thomas Henry Huxley (1825–1895) [Foster, 2002, page 82; Milner, 1999, page 90]. Huxley, in turn, was a leading defender of Charles Darwin's work on natural selection (1809–1882). Darwin and Huxley knew Lankester through their friendship with Lankester's father. In addition, Edwin Ray Lankester became a "close friend of Karl Marx in the last few years of Marx's life" [Foster, 2000] while he was a professor at University College, London. Lankester believed that humans were responsible for the extinction of species. In "The Effacement of Nature by Man," which was written before World War I and published in his collection *Science from an Easy Chair; A Second Series*, Lankester pointed out man's activities have led to "vast destruction and defacement of the living world by the uncalculating reckless procedure of both savage and civilized man." [Lankester, 1913, page 365] Furthermore, Lankester continued, "so far as we can see, if man continues to act in the reckless way which has characterized his behavior hitherto, he will multiply to such an enormous extent that only a few kinds of animals will be left on the face of the globe." [Lankester, 1913, page 366]

Lankester and Arthur Tansley (1871–1955) were among the socialists who incorporated naturalistic and ecological concepts into their thinking [Foster, 2002, pages 79–80]. Tansley was influenced by Lankester and botanist Francis Wall Olivier while attending University College, London. Lankester and Tansley were more Fabian socialist than Marxist socialist. Like Marx, Lankester and Tansley were materialists. Tansley was a leading plant ecologist and coined the term "ecosystem." He believed that human activity was destructive to the environment.

Tansley worked with Charles Elton (1900–1991) in the Nature Conservancy [Bocking, 2012]. Elton had a "fiery" writing style which he used to condemn the use of pesticides in 1958 [*The Ecology of Invasions by Animals and Plants*, 1958, pages 137–142].

Vladimir Lenin was a Marxist, revolutionary, and politician. He was a member of the Bolshevik Party, which was a wing of Russian Social-Democratic Workers Party. Lenin led the Bolshevik Revolution in 1917 and became the first leader of the Soviet state. The Bolshevik Party renamed itself the Russian Communist Party in 1918.

Lenin was aware that Marx relied on Liebig's work in agriculture [Lenin, 1901]. Lenin was concerned that private ownership of agriculture in a capitalist system could be harmful to the environment. As head of state, Lenin issued mandates that gave the state control of land, minerals, waters, forests, and natural resources. The Soviet state, under Lenin's leadership, issued two decrees in October, 1917. The first decree withdrew Russia from World War I, and the second decree entitled "On Land" brought all land under government control.

The second decree was codified by legislation from Lenin's Central Executive Committee in 1918 entitled "The Fundamental Law of Land Socialization." Part 1, Article 1 of the legislation abolished private ownership of land by asserting that "All private ownership of land, minerals, waters, forests, and natural resources within the boundaries of the Russian Federated Soviet Republic is abolished forever" [Lenin CEC, 1918]. Part 1, Article 2 of the legislation required that all land be "handed over without compensation (open or secret) to the toiling masses for their use" [Lenin CEC, 1918]. Lenin's belief that natural resources should not be used for profit is implemented in Part 1, Article 17 which stated that "Surplus income derived from the natural fertility of the soil or from nearness to market is to be turned over to the organs of the Soviet Government, which will use it for the good of society" [Lenin CEC, 1918].

19.6 *The Fabian Society: Evolutionary rather than revolutionary*

Marx and Engels supported the use of proletariat revolution to overthrow bourgeoisie rule. Lenin led a revolution in 1917 to wrest control from Czar Nicholas II and install a Marxist system in Russia. Not everyone agreed that revolution could be used in every state. A group called the Fabian Society believed that a strategy of gradual change, rather than revolutionary change, would be needed to replace stable democracies with Marxist or socialist systems.

The Fabian Society was founded in Britain in 1884. It was one of the original founders of the British Labour Party and is constitutionally affiliated to the party as a Socialist Society. Members of the Fabian Society "advocate gradualist, reformist and democratic means in a journey towards radical ends." [Fabian Society, 2017b] Former British

Prime Minister Tony Blair is a well-known member of the Fabian Society. The Fabian Society is discussed in more detail in Section 21.

19.7 The Marxist view of environmentalism goes global

Lenin, like Marx, was interested in changing the world. Tactics to achieve social progress in the political world include establishing a government that has the power to grant and limit rights, control the economy, redistribute wealth, and define morality. Former U.S. President Woodrow Wilson was a progressive who supported the formation of the League of Nations in Geneva, Switzerland after World War I (1920). The League of Nations was the first attempt of the modern era to provide a forum for resolving international conflicts. Its successor, the United Nations, was established in New York City after World War II (1945).

The United Nations was formed to promote international cooperation, and to maintain and create international order. Josef Stalin (1878–1953) served as Secretary General of the Soviet Communist Party from 1922 until his death in 1953. Stalin followed Lenin as leader of the Soviet Union after Lenin's death in 1924. The line of succession as Soviet prime minister was Lenin (1923–1924) followed by Alexei Rykov (1924–1930), Vyacheslav Molotov (1930–1941), and then Stalin (1941–1946). Stalin led the Soviet Union into the United Nations as a charter member.

The Soviet Union joined the United States, Great Britain, China and France as permanent members of the UN Security Council with veto power over any resolutions introduced in the Security Council. The Russian Federation took over the Soviet Union's role in the United Nations when the Soviet Union dissolved in 1991. As we shall see, the United Nations has been used as a platform for proponents of one world government. Figure 19-2 illustrates connections from Hegel to the United Nations.

Figure 19-2. Connections from Georg Hegel to the UN.

20. Oligarchic political obstacles

Adoption and implementation of the Goldilocks Policy for energy transition can be blocked by the ruling class. The ruling class is composed of people with the ability to unduly influence government policies. The composition of the ruling class ranges from a low-level bureaucrat with the authority to control public resources to billionaires who influence the behavior of policy-makers. The type of government that consists of a minority ruling class exercising control over the majority of people is an oligarchy.

Two oligarchic political models and their potential as obstacles to implementing the Goldilocks Policy are considered in this chapter. The first model is a plutocracy model based on Harvard historian Carroll Quigley's *Tragedy and Hope — A History of the World in Our Time* [Quigley, 1966]. The second model is the bureaucratic ruling class model based on Bruno Rizzi's *The Bureaucratization of the World* [1939, French edition; see also Goldberg, 2014]. The managerial ruling class model based on James Burnham's *The Managerial Revolution* [1941] and the Deep State ruling class model [see, for example, Lofgren, 2014; and Chaffetz, 2018] are similar to the bureaucratic ruling class model.

Quigley argued that people with wealth can control political leaders in government, regardless of their public political affiliation. Rizzi and Burnham observed that unelected and mostly unaccountable bureaucrats run modern institutions in the public and private sector. Some elected officials may be accepted as part of the ruling class if their influence aligns with the beliefs of the ruling class. Career bureaucrats can either facilitate changes in government policy when the government changes or they can choose to hinder changes. The ruling class models described in this chapter can explain why the replacement of one political party by another in countries like the United States may not lead to a major change in central government policies, including energy policies.

20.1 *Quigley's plutocracy model*

American voters have often complained that the Federal government continues business as usual even when a new political party takes power. Historian Carroll Quigley provided a possible explanation for the apparent inertia of the Federal government in his 1966 book *Tragedy and Hope — A History of the World in Our Time* [Quigley, 1966]. Quigley taught at Princeton, Harvard, and the Foreign Service School of Georgetown University and was the author of the text *The Evolution of Civilizations: An Introduction to Historical Analysis* [Quigley, 1961]. Quigley's *Tragedy and Hope* focused on the role of wealth in the political process from the mid-19th century through the mid-20th century. He was influenced by the brutality of the wars during this period, which included World War I and World War II, and the Great Depression. He believed that the tragedy of the period was a result of "materialism, selfishness, false values, hypocrisy and secret values" [Quigley, 1966, page 1310]. He believed that our hope for the future was to build a better world based on Western virtues of "generosity, compassion, cooperation, rationality, and foresight, and finding an increased role in human life for love, spirituality, charity, and self-discipline" [Quigley, 1966, page 1311].

A critical review of Quigley's work by W. Cleon Skausen entitled *The Naked Capitalist* [Skausen, 1970] hypothesized that Quigley's 1300+ page book *Tragedy and Hope* exposed activities of people who finance and significantly influence the political system. Gary Allen and Larry Abraham expressed concerns about the influence of private interests in *None Dare Call It Conspiracy* [Allen and Abraham, 1971]. The importance of money and mass media in democratic systems and public policy debates requires that we consider Quigley's views.

Quigley had studied the operation of a network of wealthy individuals for twenty years and was permitted to "examine its papers and secret records" for two years in the early 1960's [Quigley, 1966, page 950]. He said the network wanted to remain unpublicized, but Quigley argued that "its role in history is significant enough to be known" [Quigley, 1966, page 950].

History is full of examples of conquerors who sought to control others. Most conquerors are known by the geographic extent of their conquests. Quigley's thesis was that a group of people, which we call the network, was attempting to use money to control human affairs around the world.

John Ruskin and the network

The network was formed in the latter half of the 19th century. According to Quigley [Quigley, 1966, p. 130], the network was originally based on the ideas of John Ruskin (1819–1900), an Oxford University graduate who was appointed the Slade professor of art at Oxford in 1870. Ruskin believed that the state should take control of the means of production and distribution for the sake of the community. He was prepared to place control of the state in the hands of an elite ruling class, even if that meant a single person: "My continual aim has been to show the eternal superiority of some men to others, sometimes even of one man," [Clark, 1964, page 267].

The network extended its influence by attracting disciples through education and by buying support while other people were fighting for freedom and independence. Ruskin expressed his contemptuous view of freedom by describing the freedom of the house fly: "I believe we can nowhere find a better type of a perfectly free creature than in the common house fly… There is no courtesy in him; he does not care whether it is king or clown he teases; and in every pause of his resolute observation, there is one and the same expression of perfect egotism, perfect independence and self-confidence, and conviction of the world's having been made for flies… You cannot terrify him, nor govern him, nor persuade him, nor convince him. He has his own positive opinion on all matters …" [Clark, 1964, page 302]

Clark believed that "the authoritarian element in Ruskin's ideal policy … was derived directly from the source book of all dictatorships, Plato's Republic. He read Plato almost every day…" Furthermore, according to Clark, "Plato wanted a ruling class with a powerful army to keep it in power and a society completely subordinate to the monolithic authority of the rulers." [Clark, 1964, page 269]

Plato's ideal society would abolish marriage and family. All men would belong to all women and vice versa. Selective breeding would be supervised by the state. Children that were considered inferior or crippled would be destroyed and the rest would be raised by the state as soon as they were weaned. There would be no differentiation between men's roles and women's roles: they would all perform labor and they would all fight wars.

Society would consist of a hierarchy of three classes: the ruling class, the military class, and the working class. The ruling class would share property, participate in communal families, and devote their intellectual efforts to determining what was best for the lower classes.

England was still a colonial power when Ruskin brought his views to Oxford University, the epitome of English academia, during the 19th century. He taught students at Oxford as if they were members of the privileged, ruling class. Ruskin believed that the English ruling class should instill in the lower classes in England and throughout the world their values and traditions. This was not a benevolent or arrogant belief, but an attempt to prevent the much more populous lower classes from abandoning the values and traditions that were needed to preserve the status of the ruling class. Ruskin had an immediate impact on one of his first students, Cecil John Rhodes (1853–1902).

Cecil Rhodes

Cecil Rhodes was born in Britain and was present for Ruskin's inaugural lecture at Oxford [NWE-Rhodes, 2018]. Rhodes' attachment to British imperialism was reinforced by Ruskin's lecture. Although Rhodes only attended Oxford for one term in 1873 before traveling to South Africa. Rhodes returned to Oxford for a second term in 1876. His experience at Oxford helped him formulate a goal of extending British rule over the entire world. This vision required wealth.

Rhodes followed his brother Herbert to the diamond fields in Kimberley, South Africa in 1871, when he was 18. Cecil supervised work on Herbert's claim and was involved with business speculation on his brother's behalf. Rhodes was a millionaire by the time he was 20 and traveled to Oxford in 1873. He left the Kimberley diamond fields in the care of his brother and his associate Charles D. Rudd.

The diamond fields were in a depression in 1874 and 1875. Rhodes used the depression as an opportunity to increase his interest in the de Beers mine. He and Rudd launched the De Beers Mining Company in 1880. De Beers was funded by NM Rothschild & Sons in London. NM Rothschild was banker and politician Nathaniel Meyer Rothschild (1840–1915), also known as the first Baron Rothschild.

Rhodes used his wealth to support political allies in England and South Africa. He became Prime Minister of the Cape Colony from 1890 to 1896. His policies helped extend British imperial policies in South Africa. He hoped to establish a strip of British territory from the Cape of Good Hope to Egypt with the goal of building a railroad from one end of Africa to another. His adventurism led to trouble with the Boer Republic of the Transvaal. Rhodes supported an unsuccessful attack on the Transvaal in 1895 that led to Rhodes resignation as Prime Minister.

Rhodes acquired mineral concessions using his wealth and political allies, such as local representatives of the British government in Africa. His growing wealth gave him the resources to implement his primary political objective: expand the British Empire. Rhodes contended [Flint, 1974] that the British were "the finest race in the world and that the more of the world we inhabit the better it is for the human race."

Rhodes died in 1902 as one of the wealthiest men in the world. His first will was written in 1877 before Rhodes had acquired significant wealth. In it, he planned to create a secret society that would continue his efforts to bring the world under British rule [NWE-Rhodes, 2018]:

"To and for the establishment, promotion and development of a Secret Society, the true aim and object whereof shall be for the extension of British rule throughout the world, the perfecting of a system of emigration from the United Kingdom, and of colonisation by British subjects of all lands where the means of livelihood are attainable by energy, labour and enterprise, and especially the occupation by British settlers of the entire Continent of Africa, the Holy Land, the Valley of the Euphrates, the Islands of Cyprus and Candia, the whole of South America, the Islands of the Pacific not heretofore possessed by Great Britain, the whole of the Malay Archipelago, the seaboard of China and Japan, the ultimate recovery of the United States of America as an integral part of the British Empire, the inauguration of a system of Colonial representation in the Imperial Parliament which may tend to weld together the disjointed members of the Empire and, finally, the foundation of so great a Power as to render wars impossible, and promote the best interests of humanity."

In his last will and testament, Rhodes provided the resources needed to establish the Rhodes Scholarships. Students from territories under British rule, formerly under British rule, or Germany were eligible to apply for scholarships to study at Oxford University.

Cecil Rhodes and the secret society

Cecil Rhodes used his wealth to fund a Secret Society with the goal of establishing a New World Order under British rule. Quigley explained that three men met in London in 1891 to form what has become known as Rhodes Secret Society [Quigley, 1981, Chapter 1]. The three men were led by Cecil Rhodes, and included journalist William T. Stead, and Reginald Baliol Brett, who was also known as Lord Esher. Quigley described Brett as "friend and confidant of Queen Victoria, and later to be the most influential adviser of King Edward VII and King George V." [Quigley, 1981, page 3] The three men developed a plan of organization that included an inner circle known as The Society of the Elect, and an outer circle referred to as The Association of Helpers. The organization was run by the leader Rhodes and a "Junta of Three" that included Stead, Brett and Alfred Milner (1854–1925). Milner was an associate of Baron Rothschild and helped found Rhodes Secret Society.

Rhodes led the group from 1891 until his death in 1902. Stead was the most influential member while Rhodes was alive. Milner became leader of the group after Rhodes' death. The expression "Rhodes Secret Society" best describes the organization from 1891 to 1902. Under Milner's leadership, the goals of the organization remained the same but the character of the organization changed. Quigley referred to the organization as Rhodes Secret Society before 1901 and the Milner group thereafter [Quigley, 1981, page 4].

According to Quigley [Quigley, 1966, page 132]: "As governor-general and high commissioner of South Africa in the period

1897–1905, Milner recruited a group of young men, chiefly from Oxford and from Toynbee Hall, to assist him in organizing his administration. Through his influence these men were able to win influential posts in government and international finance and became the dominant influence in British imperial and foreign affairs up to 1939. Under Milner in South Africa they were known as Milner's Kindergarten until 1910. In 1909–1913 they organized semisecret groups, known as Round Table Groups, in the chief British dependencies and the United States. These still function in eight countries. They kept in touch with each other by personal correspondence and frequent visits, and through an influential quarterly magazine, The Round Table, founded in 1910 and largely supported by Sir Abe Bailey's money. In 1919 they founded the Royal Institute of International Affairs (Chatham House) for which the chief financial supporters were Sir Abe Bailey and the Astor family (owners of *The Times*). Similar Institutes of International Affairs were established in the chief British dominions and in the United States (where it is known as the Council on Foreign Relations) in the period 1919–1927. After 1925 a somewhat similar structure of organizations, known as the Institute of Pacific Relations, was set up in twelve countries holding territory in the Pacific area, the units in each British Dominion existing on an interlocking basis with the Round Table Group and the Royal Institute of International Affairs in the same country. In Canada the nucleus of this group consisted of Milner's undergraduate friends at Oxford (such as Arthur Glazebrook and George Parkin), while in South Africa and India the nucleus was made up of former members of Milner's Kindergarten."

The Royal Institute of International Affairs, also known as Chatham House, was founded in 1919 at the Paris Peace Conference that ended World War I. According to the Chatham House website [Chatham House, 2018], "In 1919 British and American delegates to the Paris Peace Conference, under the leadership of Lionel Curtis, conceived the idea of an Anglo-American Institute of foreign affairs to study international problems with a view to preventing future wars. In the event, the British Institute of International Affairs was founded separately in London in July 1920. The American delegates developed the Council on Foreign Relations in New York as a sister institute." The British Institute of International Affairs received its Royal Charter in 1926.

The League of Nations was founded on January 10, 1920 as a result of the same Paris Peace Conference that ended World War I. The League of Nations was a forerunner of the United Nations and was formed as an international organization that would provide a forum for promoting international cooperation and achieving peace and security. United States President Woodrow Wilson is credited with founding the League of Nations.

The United States affiliate of the Round Table, the Council on Foreign Relations (CFR), was founded in 1921. The CFR specialized in foreign affairs and international relations. It participated in the 1945 United Nations Conference on International Organization in San Francisco. The Charter of the United Nations was signed at the conclusion of the conference. Figure 20-1 highlights the connections from John Ruskin to the United Nations (UN).

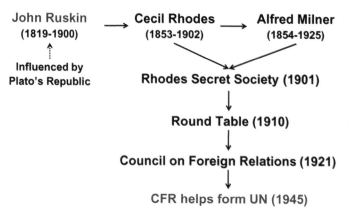

Figure 20-1.   Connections from John Ruskin to the UN.

20.2 *The bureaucratic ruling class model*

In *The Bureaucratization of the World* [Rizzi, 1939], Communist intellectual Bruno Rizzi argued that the Soviet Union, as ruled by Joseph Stalin, was not a Communist system, but a new kind of system that he called "bureaucratic collectivism." He believed the Soviets had replaced capitalist and aristocratic ruling classes with a new ruling class consisting of economic and political bureaucrats. The bureaucrats retain power by controlling the armed forces, both police and military, in a police state. In Section VIII of his book, Rizzi concluded that defense of the Soviet Union meant defending a new system of exploitation which was "imposing itself on the entire world." He went on to say that "the Stalinist regime is intermediate, it throws aside outdated capitalism but it does not rule out socialism for the future. It is a new social form, based on class property and class exploitation."

Bolshevik theoretician Leon Trotsky rejected Rizzi's arguments. In referring to Bruno Rizzi as Bruno R., Trotsky analyzed Rizzi's Theory of Bureaucratic Collectivism in "The USSR in War" [Trotsky, 1939]. Trotsky did not agree that the Stalinist bureaucracy should be considered a new class. He did recognize that the Stalinist regime was either "an abhorrent relapse in the process of transforming bourgeois society into a socialist society, or the Stalinist regime is the first stage of a new exploiting society." If the latter option is correct, then "the bureaucracy will become a new exploiting class." [Trotsky, 1939, page 6]

James Burnham challenged Trotsky's critique of Rizzi's bureaucratic collectivism in *The Managerial Revolution* [Burnham, 1941]. He believed that a new ruling class was emerging as part of a social revolution. Burnham identified a social revolution as a revolution that consisted of three constituents [Burnham, 1941, page 5]: a drastic change occurs in the most important social (economic and political) institutions; concurrent with change in social institutions is a shift in the cultural institutions and dominant beliefs about the role of humanity in our world and the universe; and a change in the group of people who control the greater part of power and privilege in society.

Burnham hypothesized that he lived in a period of social revolution, which was a period of transition from one type of society to another [Burnham, 1941, page 7]. With these observations, Burnham presented his theory of the managerial revolution to explain the period of transition and predict the type of society that would emerge. He believed that the transition was from a type of society called capitalist or bourgeois to a society he called managerial [Burnham, 1941, page 71]. Burnham said that a power struggle was occurring during the period of transition as the emerging class of managers drove for social dominance. He believed that the struggle was occurring worldwide, and that the managerial class would become the ruling class. The managerial class would establish social dominance as the ruling class by using the state to control the major instruments of production [Burnham, 1941, page 72].

Burnham made a distinction between control and ownership [Burnham, 1941, Chapter VII]. The managerial class will be the group of people who control the major instruments of production. The group of people that owned the major instruments of production under capitalism will eventually cede ownership to the people in control.

Trotsky continued his arguments against the theories advocated by Rizzi, Burnham, and their allies in his book *In Defense of Marxism* [Trotsky, 1942]. Trotsky believed that Burnham was anti-dialectic, that is, that Burnham did not use the dialectical process to seek a synthesis of different points of view represented by a thesis and its antithesis. Instead, Trotsky accused Burnham of being a pragmatist. Consequently, Burnham's reasoning and conclusions had to be incorrect since he was not properly applying dialectics.

The key difference between Trotsky and Burnham was their view of Stalin's totalitarian government. Trotsky viewed Stalin's government as a government in transition from a bourgeoisie society to an egalitarian society. By contrast, Burnham viewed it as a new version of oppression. The point of view expressed by Burnham and his allies was fictionalized by socialist author George Orwell in *Animal Farm* [Orwell, 1945].

Managerialism

We previously discussed a third vision proposed by Burnham [1941] which we refer to as managerialism. Burnham believed that capitalism was disappearing, but he did not believe that socialism was taking its place. Instead, Burnham argued that a managerial revolution was occurring. George Orwell (actual name Eric Blair, 1903–1950) provided a succinct summary of Burnham's view [Orwell, 1946]: "Capitalism is disappearing, but Socialism is not replacing it. What is now arising is a new kind of planned, centralized society which will be neither capitalist nor, in any accepted sense of the word, democratic. The rulers of this new society will be the people who effectively control the means of production: that is, business executives, technicians, bureaucrats and soldiers, lumped together by Burnham, under the name of "managers". These people will eliminate the old capitalist class, crush the working class, and so organize society that all power and economic privilege remain in their own hands. Private property rights will be abolished, but common ownership will not be established. The new "managerial" societies will not consist of a patchwork of small, independent states, but of great super-states grouped round the main industrial centers in Europe, Asia, and America. These super-states will fight among themselves for possession of the remaining uncaptured portions of the earth, but will probably be unable to conquer one another completely. Internally, each society will be hierarchical, with an aristocracy of talent at the top and a mass of semi-slaves at the bottom."

Some believe that Orwell's classic dystopian novel *1984* [Orwell, 1949], published a century after the founding of the Fabian Society in 1884, was a cautionary tale about life in a society based on Fabian Socialism. Orwell was a socialist who had concerns about the socialism advocated by socialists that were part of the bourgeois rather than the working class. Orwell expressed his political views in his 1937 book *The Road to Wigan Pier* [Orwell, 1937, pages 161–162]: "Sometimes I look at a Socialist — the intellectual, tract-writing type of Socialist, with his pullover, his fuzzy hair, and his Marxian quotation — and wonder what the devil his motive is. It is often difficult to believe that it is a love of anybody, especially of the working class, from whom he is of all people the furthest removed. The underlying motive of many Socialists, I believe, is simply a hypertrophied sense of order. The present state of affairs offends them not because it causes misery, still less because it makes freedom impossible, but because it is untidy; what they desire, basically, is to reduce the world to something resembling a chessboard. Take the plays of a lifelong Socialist like [George Bernard] Shaw. How much understanding or even awareness of working-class life do they display?" Orwell concluded: "The truth is that, to many people calling themselves Socialists, revolution does not mean a movement of the masses with which they hope to associate themselves; it means a set of reforms which 'we', the clever ones, are going to impose upon 'them', the Lower Orders." [Orwell, 1937, page 162]

The deep state

The Deep State refers to individuals and institutions that exercise political power independent of, and sometimes in opposition to, civilian political leaders. The term was traditionally applied to developing countries where top military and government officials exercised political power in nominally democratic societies and replaced elected leaders when they deemed it worthwhile. In his farewell address on January 17, 1961, United States

President Dwight Eisenhower spoke of the military-industrial complex as the "conjunction of an immense military establishment and a large arms industry" [Eisenhower, 1961]. Eisenhower warned that "In the councils of government, we must guard against the acquisition of unwarranted influence, whether sought or unsought, by the military-industrial complex. The potential for the disastrous rise of misplaced power exists and will persist." [Eisenhower, 1961]

A security and intelligence apparatus arose in the United States after the September 11, 2001 terrorist attacks. The apparatus was obscure to public scrutiny and appeared to be unaccountable to elected officials and the civilian legal system. Some Americans began to express concerns that a Deep State may be operating within the United States.

Long-time Congressional staff member Mike Lofgren provided an analysis of the Deep State in the United States based on 28 years of service as a Congressional staff member, top security clearance, and specialization in national security. According to Lofgren [Lofgren, 2014], "All complex societies have an establishment, a social network committed to its own enrichment and perpetuation." The American Deep State is unique in the world because of its scope, financial resources, and global reach. Lofgren wrote that the American Deep State "is a hybrid of national security and law enforcement agencies: the Department of Defense, the Department of State, the Department of Homeland Security, the Central Intelligence Agency and the Justice Department. I also include the Department of the Treasury because of its jurisdiction over financial flows, its enforcement of international sanctions and its organic symbiosis with Wall Street." Lofgren believed that these agencies were coordinated by the Executive Office of President Obama via the National Security Council. In addition to agencies of the Executive Branch of the United States government, Lofgren said that some key areas of the judiciary belong to the Deep State, and mentioned the Foreign Intelligence Surveillance Court (the FISA court) as an example. The final government component of the Deep State consisted of some congressional leaders and members of the defense and intelligence committees. Former Utah Congressman Jason Chaffetz presented a similar view of the Deep State based on his experience in Congress during the Obama Administration [Chaffetz, 2018]. If Lofgren's analysis is correct, or nearly so, the Deep State had to adapt to a change in leadership in the Executive Branch when Donald Trump became President in January 2017.

21. Fabian socialism as a political obstacle

The Fabian Society is a British socialist organization that was founded in London in 1884. The purpose of the Fabian Society was to gradually usher in a one world collectivist state. Fabian Socialists seek to advance the principles of democratic socialism by gradually introducing and adopting reforms in democracies. They believe that the gradual transition to democratic socialism would be more effective than attempting to invoke the Marxist doctrine of revolutionary overthrow. If successful, Fabian Socialism would usher in a one-world globalist state using gradual reforms in democratic systems. The impact of Fabian Socialism and its support of globalism on energy policy are discussed here.

21.1 *The Fabian Society*

The Fabian Society derived its name from Roman general Quintus Fabius Maximus Verrucosus (275–203 B.C.), who received the surname Cunctator, or Lingerer, by delaying attacks on Hannibal's invading Carthaginian army until the time was right. The first Fabian pamphlet explained its name with the note: "For the right moment you must wait, as Fabius did most patiently, when warring against Hannibal, though many censured his delays; but when the time comes you must strike hard, as Fabius did, or your waiting will be in vain, and fruitless." [Fabian Society, 2017a] The Fabian Society was founded in 1884 as an offshoot of Thomas Davidson's Vita Nuova or

Fellowship of the New Life [Pease, 1916]. Davidson (1840–1900) was a Scottish-American philosopher and lecturer. He founded the Fellowship of the New Life in London in 1883, and later in New York.

Blakewell wrote a tribute to Davidson after his death [Blakewell, 1901]. Blakewell wrote that Davidson believed that "the times were religiously and socially out of joint, and that it was his duty, as it was that of every man, to do his best to set them right. With this end, he took an active interest in the founding of the London Fabian Society." [Blakewell, 1901, page 447] The founding goals of the organization were to gather and disseminate information that might lead to an improvement in social conditions. Blakewell pointed out that the organization did not begin as a socialist organization, and Davidson lost interest in the organization as it drifted toward socialism. Davidson believed that social reform should begin with the individual, and especially with the education of the individual. Consequently, according to Edward Pease, who served as Fabian Society secretary and historian, Davidson "was the occasion rather than the cause of the founding of the Fabian Society. His socialism was ethical and individual rather than economic and political." [Pease, 1916, Chapter 1]

Founding members of the Fabian Society included Pease, Edith Nesbith, Hubert Bland, and Frank Podmore. Some notable early members of the group included Irish playwright George Bernard Shaw (1856–1950), Sydney Webb (husband of author Beatrice Potter who later became a Fabian Society member), and Eleanor Marx (eldest daughter of Karl Marx). Author H.G. Wells was a member of the Fabian Society for a short time (1903–1908). Wells did not stay with the Fabian Society very long because of a controversy that brought into focus the direction of the group in the early 1900s.

According to Pease [Pease, 1916, Chapter IX], the controversy began when Wells presented a paper entitled "Faults of the Fabians" to the Fabian Society in February, 1906. Wells suggested that the Fabian Society was being too timid to achieve its goal of restructuring society and proposed a more activist agenda. Later that year, Scottish socialist Keir Hardie and the Independent Labour Party made Fabian Society members reconsider the political landscape in Britain.

Keir Hardie was the first socialist member of the British House of Commons in 1892. He helped establish the Independent Labour Party in 1893. In doing so, Hardie lost the support of the Liberal Party and subsequently lost his seat in Parliament. In 1900, the three socialist organizations in Britain — the Independent Labour Party (ILP), the Social Democratic Federation (SDF), and the Fabian Society — joined forces with labor unions to form the Labour Representation Committee (LRC) with Ramsay MacDonald as secretary. MacDonald was a member of both the Fabian Society and the ILP. The LRC consisted of two members from the ILP, two members of the SDF, one member from the Fabian Society, and seven members of trade unions. Hardie was one of the members from the ILP and Pease was the member from the Fabian Society. Hardie was one of two ILP members that were elected to the House of Commons in 1900. In 1906, ILP candidates, including Hardie and MacDonald, won 29 seats in the House of Commons. The group met that same year and decided to change from the LRC to the Labour Party. Hardie was elected as Labour Party Chair and MacDonald was selected to serve as Labour Party secretary [Simkin, 2017]. Ramsay MacDonald later became the first British Prime Minister from the Labour Party in 1924.

The election success of the ILP in 1906 showed the Fabian Society that socialism was becoming a political force in Britain. Pease observed that Wells' "proposal for an enlarged and invigorated society came at the precise moment" when the realization that the Wells project was possible [Pease, 1916, Chapter IX]. As of 2018, all British Prime Ministers that were members of the Labour Party (UK) were also members of the Fabian Society. They are Ramsay MacDonald, Clement Atlee, Harold Wilson, James Callaghan, Tony Blair and Gordon Brown. This history

Obstacles to Adopting the Goldilocks Policy | 123

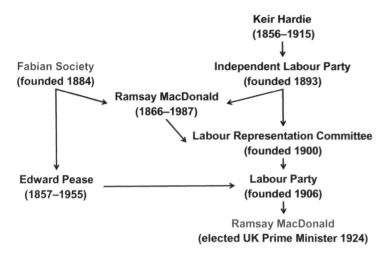

Figure 21-1. Connections from the Fabian Society to National Governance.

established connections from the Fabian Society to national governance, as illustrated in Figure 21-1.

The Executive Committee of the Fabian Society considered Wells' 1906 proposal and a counter-proposal written by Shaw was submitted to the membership. Pease believed that "the real issue was a personal one…Was the Society to be controlled by those who had made it or was it to be handed over to Mr. Wells? We knew by this time that he was a masterful person, very fond of his own way, very uncertain what that way was, and quite unaware whither it necessarily led. In any position except that of leader Mr. Wells was invaluable, as long as he kept it! As leader we felt he would be impossible, and if he had won the fight he would have justly claimed a mandate to manage the Society on the lines he had laid down. As Bernard Shaw led for the Executive, the controversy was really narrowed into Wells versus Shaw." [Pease, 1916, Chapter IX]

Wells did serve on the Executive Committee until his resignation in 1908. Pease reported that "Mr. Wells was the spur which goaded us on, and though at the time we were often forced to resent his want of tact, his difficult public manners, and his constant shiftings of policy, we recognized then, and we remember still, how much of permanent value he achieved." [Pease, 1916, Chapter IX] According to Pease, Wells chief achievements were his books. Pease singled out Wells' book *New Worlds for Old* [Wells, 1909] which was published by the Fabian Society in 1908 while Wells was still a member. It was published commercially in 1909 by The MacMillan Company in New York. Pease called *New Worlds for Old* "perhaps the best recent book on English Socialism." [Pease, 1916, Chapter IX]

21.2 *Fabian globalism*

The Fabian Society supports global governance by a socialist government. Leonard S. Woolf, husband of author Virginia Woolf, published two reports prepared for the Fabian research department and the Fabian Committee for a Supranational Authority that will Prevent War in 1916 during World War I. Woolf's reports were published as a book [Woolf, 1916] with an introduction by Fabian Society member Bernard Shaw. In the introduction, Shaw wrote that "Unless and until Europe is provided with a new organ for supranational action, provided with an effective police force, all talk of making an end of war is mere waste of breath" [Woolf, 1916, page xv]. Woolf wrote that "The alternative to war is law. What we have to do is to find some way of deciding differences between States, and of securing the same acquiescence in the decision as is now shown by individual citizens in a legal judgment. This involves the establishment of a Supranational Authority, which is the essence of our proposals" [Woolf, 1916, page 372]. Figure 21-2 illustrates the connections from the Fabian Society to Globalism in the form of a supranational authority.

A set of articles for establishing a supranational authority was presented by Woolf [Woolf, 1916, Section II]. The first step in establishing

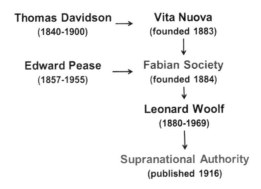

Figure 21-2. Connections from the Fabian Society to Globalism.

a supranational organization was to form an International High Court. The supranational League of Nations was established three years later in 1919 following the conclusion of World War I. The League of Nations formed the Permanent Court of International Justice in The Hague, Netherlands. The Court was known as the World Court and existed from 1922 to 1946.

Some people credit United States President Woodrow Wilson with initially conceiving of The League of Nations. Wilson was a major proponent of The League of Nations. Despite Wilson's support, The League of Nations failed in the 1930s, in part, because the United States did not become a member. Many Americans had not supported United States involvement in World War I and did not want to get entangled in European affairs after the war ended. When The League of Nations failed, a second supranational authority was established. The United Nations was formed in 1945 after World War II with the support of the Council on Foreign Relations, an offshoot of the Milner Group, and the Fabian Society. The World Court became the International Court of Justice in 1945 and was authorized by the United Nations Charter. Connections from the Fabian Society to the United Nations are outlined in Figure 21-3.

Figure 21-3. Connections from the Fabian Society to the United Nations.

22. Globalism as a political obstacle

Brooks highlighted a battle between two competing and irreconcilable visions of globalism in the United States [Brooks, 2010, page 1]. He argued that the people of the United States are in a political struggle between free enterprise and European-style statism. Brooks defined free enterprise as "the system of values and laws that respects private property, encourages industry, celebrates liberty, limits government, and creates individual opportunity." [Brooks, 2010, page 3] By contrast, European-style statism is the democratic socialism preferred by Fabian Socialists. The implementation of democratic socialism in Europe relies on expanded government bureaucracies, government control of the private sector, and increasing income distribution. Fabian Socialists have the goal of establishing their concept of a supranational authority as the global government. The history of global governance is presented here.

22.1 *World War I and The Inquiry*

Woodrow Wilson has been credited with founding the League of Nations as the first supranational authority of the modern age. The League of Nations was the last point of Wilson's 14-point plan to end World War

I, and the first modern attempt to establish global governance. We need to review the history of the establishment of the League of Nations to understand the intended role of the organization and why the United States decided not be become a member of the League of Nations.

Wilson won a 3-way race for United States President in 1912. He defeated then current President William Howard Taft and former President Theodore (Teddy) Roosevelt. Teddy Roosevelt served as President from 1901–1909, followed by Taft from 1909–1913, and then Wilson from 1913–1921. World War I started in 1914 during Wilson's first term.

President Woodrow Wilson kept the United States out of World War I, even after the British passenger ship Lusitania was sunk by a German submarine in May, 1915. The attack killed both British and American passengers. Germany declared the waters around the British Isles a war zone and announced that they would attack any shipping in the area. Wilson knew the American public was not ready to participate in what many considered to be a European war. When Wilson ran for his second term in 1916, he based his campaign on the anti-war slogan "he kept us out of war" and legislative achievements such as prohibition of child labor and limiting the work day of railroad workers to 8 hours.

Once re-elected, Wilson used the German policy of unrestricted submarine warfare as a reason to enter the war on the side of the Allies led by Britain and France. On April 2, 1917 Wilson asked Congress to declare war on Germany. It was clear the Allies were going to win by December, 1917 when Wilson appointed a committee of experts to prepare specific recommendations for a comprehensive peace settlement. The committee was known as The Inquiry and was led by Wilson's adviser Edward M. House (1858–1938).

The Inquiry was a committee of over 100 researchers, executives and staff assistants. Many of the researchers were academics in such disciplines as geography, political science, history, economics, and international law. House appointed Walter Lippmann, then assistant to Wilson's

Secretary of War, to serve as secretary of The Inquiry. Lippmann acted as a coordinator of The Inquiry and became one of the key architects of Wilson's plan to end World War I.

22.2 *Walter Lippmann's brush with Fabian socialism*

Walter Lippmann entered Harvard in 1906 where he studied philosophy and languages (French and German). Lippmann served as president of the Intercollegiate Socialist Society (ISS) club at Harvard in 1909 [Shafer and Snow, 1962]. The purpose of the ISS was to promote socialism on college campuses. Among the founders of the ISS in 1905 were authors Upton Sinclair and Jack London. Both men were in their twenties at the time, and London was the first president of the ISS.

Lippmann completed his Bachelor of Arts degree requirements in three years and spent a fourth year at Harvard working as an assistant to philosopher George Santayana. While at Harvard, Lippmann attended a discussion course led by Graham Wallas (1858–1932), British visiting lecturer and former Fabian Socialist. The relationship between Lippmann and Wallas establishes a connection between Lippmann and Fabian Socialism.

Wallas joined the Fabian Society in 1886 along with his friend Sidney Webb. Webb became a leader of the Fabian Society and married Beatrice Potter in 1892. Sidney and Beatrice Webb founded the London School of Economics and Political Science (LSE) in 1895. The purpose of the LSE was to teach economics along more socialist lines. The Webbs asked Wallas to become the first director of the LSE. Webb turned down the offer but agreed to teach at the LSE where he became a Professor of Politics. Wallas eventually grew impatient with the unending discussions of the Fabian Society and believed the leaders of the Fabian Society had adopted anti-liberal attitudes. Wallas resigned from the executive committee in 1895 and from the Society itself in 1904. Wallas was impressed enough by Lippmann that Wallas dedicated his 1914 book entitled 'The Great Society' to Lippmann.

Lippmann formally graduated from Harvard in 1910 and began his career in journalism as a cub writer for the small left-of-center journal *The Boston Common* where he worked for liberal socialist Ralph Albertson. Lippmann was later hired by Lincoln Steffens, a major byline writer for the national magazine called *Everybody's Magazine* [Duffy 2009; Whitman, 1974].

Steffens and Lippmann supported Theodore Roosevelt's Progressive Party in the 1912 Presidential election when Roosevelt ran against Taft and Wilson. Lippmann left *Everybody's Magazine* to become executive secretary for George R. Lunn, the newly elected socialist mayor of Schenectady, New York. Lippmann left the job after four months.

Lippmann published the book *A Preface to Politics* in 1913 and co-founded *The New Republic* with friend and journalist Herbert Croly (1869–1930). Like Lippmann, Croly was educated at Harvard and became a leader of the progressive movement. Croly authored the 1909 book 'The Promise of American Life' in which he argued for greater economic

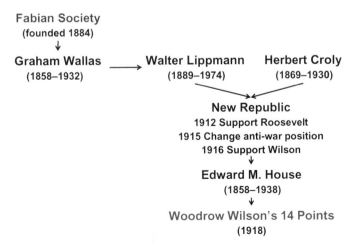

Figure 22-1. Connections from the Fabian Society to Wilson's 14 Points.

planning, more spending on education, and establishment of a society based on "the brotherhood of mankind."

Heiress Dorothy Payne Whitney and her husband, banker and diplomat Williard Straight, asked Croly to join them in launching *The New Republic* as a liberal journal that would provide intelligent liberal commentary on politics, foreign affairs and culture. Whitney and Straight agreed to fund the journal. Croly recruited Lippmann to serve on the editorial board of *The New Republic* with Croly effectively serving as Editor-in-Chief. *The New Republic* published its first article in November, 1914 [History.com Staff, 2007].

The New Republic began to change its anti-war position in 1915. Lippmann began to move away from socialism in his 1916 book *Drift and Mastery*, but retained progressive ideas. He used *The New Republic* to support Wilson's re-election, which brought him into contact with Wilson's friend and close adviser Edward M. House. House convinced Lippmann, a pacifist, to support Wilson's policy of limited military preparedness for entering the war. In 1917, Lippmann accepted an appointment to serve as assistant to Wilson's Secretary of War Nelson Baker. Within the year, Lippmann began to work as secretary of The Inquiry.

22.3 Wilson's 14 Points and the League of Nations

United States President Woodrow Wilson presented his 14 Points to Congress on January 18, 1918 [Wilson 14, 1918]. Wilson said that the "day of conquest and aggrandizement is gone by; so is also the day of secret covenants entered into in the interest of particular governments." He pointed out that the United States "entered this war because violations of right had occurred which touched us to the quick and made the life of our own people impossible unless they were corrected and the world secure once for all against their recurrence." He believed that the world should "be made safe for every peace-loving nation which, like our own, wishes to live its own life, determine its own institutions, be assured of justice and fair dealing by the other peoples of the world as against force and selfish aggression. All the peoples of the world are in effect partners in this interest..." With this preamble, Wilson introduced his 14 points as the only possible program that could ensure world peace.

Points 1 through 5 addressed international agreements, freedom of navigation of the seas, international trade, national armaments, and the free, open-mined, and absolutely impartial adjustment of colonial claims. Points 6 through 13 addressed territorial claims in many countries, including Russia, Belgium, France, Italy, Austria-Hungary, Rumania, Serbia, Montenegro, the Balkan States, Turkey, and Poland. A supranational authority was proposed in Point 14: "A general association of nations must be formed under specific covenants for the purpose of affording mutual guarantees of political independence and territorial integrity to great and small states alike."

The German government arranged a general armistice in October, 1918 along the lines of Wilson's 14-point program. Germany signed the Armistice in November, 1918.

Wilson traveled to the Paris Peace Conference in 1919. The Allies rejected many elements of Wilson's 14-point program. Instead, they demanded compensation from Germany for war damage and territorial agreements that reduced the size of Germany. The Treaty of Versailles was signed in the Palace of Versailles on June 28, 1919. The treaty included the Covenant for the League of Nations, the establishment of the Permanent Court of International Justice, and the International Labor Organization.

Walter Lippmann attended the Paris Peace Conference as a United States delegate and did not approve of the changes that were made to Wilson's 14-point program. Herbert Croly and Lippmann were critical of Wilson because they did not support the peace treaty that emerged from the Paris Peace Conference. The New Republic began to distance itself from Wilson and was used to urge opposition to the Treaty of Versailles and United States participation in the League of Nations.

Lippmann knew that the role of socialism in the world would be a factor in the League of Nations. In 1919, Lippmann wrote [Lippmann, 1919, page 63] "It will be difficult enough in all conscience to secure harmony in a League when half the world is socialist and the other half anti-socialist." Lippmann was concerned that the League would be irrelevant [Lippmann, 1919, page 64]: "…if the League is not to find itself marooned on the dry sands of irrelevance it should take steps to introduce into its own structure the conciliatory influence of the opposition parties."

The Treaty of Versailles took effect on January 10, 1920. The Treaty included demands by the Allies that were not part of Wilson's 14-point program. In March, the United States Senate rejected the Treaty of Versailles. The United States did not sign the Treaty of Versailles and did not become a member of the League of Nations.

The punitive character of the Treaty of Versailles, especially German reparation and loss of territory claimed by Germany, and the inability of the League of Nations to attract broad global support were important factors that set the stage for the rise of Adolph Hitler and the National Socialists (Nazis) in 1930s Germany. The world would have to wait until the end of World War II to see the rise of the second supranational authority established in the 20th century, the United Nations.

23. Environmentalism and the United Nations

A tactic for achieving social progress was a plan proposed by Sweden in 1967 to hold an international environmental conference. The call for an international environmental conference was supported by the Secretary-General of the United Nations, U Thant of Burma, in 1969.

U Thant believed that the United Nations and the environment could be used to implement an agenda to achieve social progress. The UN was led by the developed nations of Europe and the United States from its inception after World War II until November 1961 when U Thant assumed the post of acting secretary general. The Cold War between the blocs led by the Soviet Union and the United States was the principal focus of the UN prior to U Thant's tenure as secretary general.

U Thant viewed third world nations as a collection of non-aligned nations that could politically counter the two Cold War blocs. U Thant was from the third world. He was born in the small town of Pantanaw, Burma on January 22, 1909 and became an educator in his home town. U Thant was named Burma's representative to the UN, an ambassadorial rank, in 1957 and retained the position until he assumed the post of acting secretary general in 1961. U Thant's selection as Secretary-General of the UN signaled the arrival of the third world as a political force in the organization.

Under U Thant's leadership, the UN became less pre-occupied with the Cold War and began to consider the north-south divide between developed and developing nations. U Thant used his position as Secretary-General to encourage developed countries to share the planet's wealth, relinquish their colonial possessions, and grow the UN budget so that the UN could fund all of the programs its members were voting to support.

U Thant was frustrated by the limitations of his position: for example, he could cite UN resolutions opposing colonialism but could not force the removal of a colonial power; he could call for the redistribution of global wealth but he could not tax member states; and he could call for the use of peacekeeping forces, but he could not compel member states to provide the resources needed to keep the peace. An opportunity to acquire more global influence was provided by the public's interest in the environment.

Rachel Carson was among the first to alarm the public about the potential adverse impact of human activity on the environment. In her book Silent Spring [1962], Carson publicized the negative effects of chemical pesticides. Growing public concern about the impact of human activity on the environment in the 1960s led the General Assembly to recommend in 1968 that the UN begin collecting data on the condition of the environment worldwide and suggest protective measures [Mische and Ribeiro, 1998].

U Thant hoped that the threat of pollution worldwide would unify member states in a way that peacekeeping and economic efforts had not. U Thant told the UN General Assembly in May 1969 that the world had 10 years to avert environmental disaster. A month later he blamed the United States for most of the problems [Lewis, 1985]. In response to growing public concern, the United States, Under President Richard Nixon, established the Environmental Protection Agency in 1970.

U Thant stressed the global nature of the threat in the introduction to his 1969 report entitled "Man and His Environment" [UN, 1969] and he discussed the problems of the human environment and the need for a new world order in a 1970 paper [U Thant, 1970]. U Thant pointed out in his 1970 paper that, for the first time in history, mankind faced "not merely a threat, but an actual worldwide crisis involving all living creatures, all vegetable life, the entire system in which we live, and all nations large or small, advanced or developing." He believed that we now "face a rapidly increasing imbalance between the life-sustaining systems of the Earth and the demands, industrial, agricultural, technological, and demographic, which its inhabitants put upon it." [U Thant, 1970, page 13]

U Thant listed population growth and urbanization as two key causes of the problems of the human environment. He believed that the "unthinking exploitation and abuse of the world's natural resources, and the plunder, befouling, and destruction of our native Earth, have already gone too far for us to rely any more on pious hopes, belated promises, and tardy efforts at self-discipline." [U Thant, 1970, page 16] U Thant argued that we could only take timely and effective measures to solve the problems of the human environment by establishing a global authority that had the power to police and enforce its decisions. Furthermore, the global authority should be closely associated with the UN.

U Thant called for globalism by asking: "Do the sovereign nations of the world have the courage and the vision to set up and support such an agency now, and thus, in the interest of future generations of life on Earth, depart radically from the hitherto sacred paths of national sovereignty?" [U Thant, 1970, page 16] In addition to globalism, U Thant believed that the task of saving the environment required "nothing less than a new step toward world order would do." [U Thant, 1970, page 17] He suggested that the "crisis of the environment could be the challenge which might show us the way forward to a responsible and a just world society". He asked: "Is it unrealistic to suggest that the undoubted global challenge we now face might become the basis for a new start in world order and a more civilized and generous way of life for the peoples of the Earth?" [U Thant, 1970, page 17]

Under U Thant's leadership, the United Nations Educational, Scientific, and Cultural Organization (UNESCO) convened regional symposia on the environment in 1969 and 1971, and a world conference on the environment in 1972 [Mische and Ribeiro, 1998]. The world conference was held in Stockholm, Sweden and became known as the 1972 Stockholm Conference. The 1972 Stockholm Conference was chaired by Maurice Strong (1929–2015), who became a key player in linking environmentalism to one world government.

24. Maurice Strong and global socialist environmentalism

Maurice Strong was born in rural Oak Lake, Manitoba, Canada in 1929 during the Great Depression. He was a self-professed socialist, and told the Canadian magazine Maclean's in 1976 that "I am a socialist in ideology, a capitalist in methodology" [Bailey, 1997]. He dropped out of school in 1944 at age 14 to work as a deck hand on ships. In 1945, at age 16, he worked as a fur buyer for Hudson's Bay Company in the Canadian north. While in the north, Strong spent some time with the

Inuit Eskimos. He met Bill Richardson, who was married to Mary McColl. McColl was a member of the family behind the Canadian oil company McColl-Frontenac.

Richardson helped Strong make important contacts including Noah Monod, Treasurer of the United Nations, in 1947. Monod helped get him a job as a low-level security officer at the United Nations headquarters in Lake Success, New York, and let Strong stay for a short time at his New York apartment where he was introduced to young David Rockefeller (1915–2017). David Rockefeller was a grandson and heir of John Davison Rockefeller (1839–1937), founder of Standard Oil. David had been assigned by his employer Chase Bank to handle the United Nations account. Strong said that he "had a long and cordial relationship with David in later years." [Strong, 2000, page 73]

Strong quit his security job and moved to Winnipeg, capital of Manitoba, to work as a securities analyst a few years later. In 1951, Strong went to work for oilman John (Jack) Gallagher at Dome Petroleum. Gallagher was a veteran of Standard Oil of New Jersey. He gave Strong "the opportunity of learning the business from a more operational point of view and as the company, Dome Petroleum, grew, Strong occupied several key roles, including Vice President, Finance" [Manitou-Strong, 2017].

In the early 1950s, Strong changed jobs and traveled overseas where he scouted service station sites in east Africa for Caltex, an oil company that began as a joint venture between the Texas Company (later named Texaco) and Standard Oil of California (later named Chevron). While in east Africa he learned about the YMCA organization in Nairobi, Kenya before returning to Canada in 1954.

Strong rejoined Dome in 1955 where he profited from stock options [Roberts, 2015]. Strong became a YMCA volunteer in its World Service Program and attended the 100th anniversary of the YMCA in Paris in 1955. While in Europe, Strong visited Geneva to meet his distant American cousins. According to the Manitou

Foundation, which was founded by Maurice Strong and his second wife Hanne Marstrand in 1988, "Strong met Tracy Strong, who was the Secretary General of the World Alliance headquartered in Geneva, Switzerland and a brother of Anna Louise Strong, the American journalist whose letters from China had been such a source of Strong's early interest in China." [Manitou-Strong, 2017] The New York Times obituary said that "Tracy Strong confirmed that he and Strong did indeed have a family relationship, though somewhat distant. Strong was pleased to meet, too, his son, Robbins, of the World Council of Churches in Geneva."

Maurice Strong's relative Anna Louise Strong (1885–1970) was educated in America and traveled extensively in the communist world, notably the USSR, eastern Europe, and China. She became "an enthusiastic supporter of the Russian experiment in communism" [EB-AL Strong, 2018] and was a member of the Comintern (Communist International). Anna Louise Strong wrote several books about her travels in the communist world. "She was a close friend of Mao Zedong, whom she had first interviewed in a cave in Yenan province in 1946" [EB-AL Strong, 2018]. She settled in China in 1958. Maurice Strong's relationship to Anna Louise Strong created a special connection between Maurice and China.

Maurice Strong was appointed to the International Committee of the Canadian YMCA in 1958. Strong was familiar with the YMCA from his travels in east Africa and his connection to the YMCA gave him contacts that helped him later in life.

Strong was running an oil company by the time he was age 31. Strong's work in the oil patch included serving as president of the Canadian Industrial Gas and Power Corporation of Canada, chairman of Petro-Canada and of the Canadian Development Investment Corporation. According to the New York Times [Roberts, 2015], Strong later confessed that he had been "an environmental sinner." Strong said that "we were running the Earth without a depreciation account, in effect spending our capital."

Strong's work on corporate boards and international affairs brought him to the attention of the Minister of External Affairs, Paul Martin Senior, and Canadian Prime Minister Lester Pearson. Pearson invited Strong to join the Canadian government as a Deputy Minister with responsibility for a program known as External Aid, and which became the Canadian International Development Agency (CIDA) under Strong's leadership. His work at CIDA allowed him to return to the United Nations as a Canadian delegate [Manitou-Strong, 2017].

According to the Manitou Foundation, "Strong's work with CIDA gave him new insights into the complexities of development. He was troubled by the environmental and social disruption caused by major infrastructure projects, which CIDA supported. It wasn't long before he became involved with environmental politics" [Manitou-Strong, 2017].

In 1969, the UN General Assembly decided to convene the UN Conference on the Human Environment. This was the first major international conference on environmental issues.

The New York Times [Roberts, 2015] reported that Strong's "success in increasing foreign aid brought him to the attention of U Thant, the secretary general of the United Nations at the time, who selected him to convene the 1972 Stockholm conference." Strong's appointment required Canadian government approval, which was provided by Canada's new Liberal Party Prime Minister, Pierre Elliott Trudeau. Strong went to New York to work as both Secretary General of the 1972 Stockholm Conference and Undersecretary General of the United Nations responsible for environmental affairs. Strong used his "consummate diplomatic skills to obtain the support of the developing countries, who were extremely skeptical about environmental issues" [Manitou-Strong, 2017].

"The 1972 Stockholm Conference 'adopted a Declaration of Principles and Action Plan to deal with global environmental issues. It put the environment issue on the international agenda and confirmed its close link with development. The Stockholm Conference ... launched a new era of international environmental diplomacy'" [Manitou-Strong, 2017].

In December 1972, the UN General Assembly established the United Nations Environment Program (UNEP). Strong "was elected by the UN General Assembly to become UNEPs first Executive Director at its new headquarters in Nairobi, Kenya" [UNEP-Strong, 2018]. UNEP became the first UN agency to be headquartered in a developing country."

Strong's opening statement at the 1972 Stockholm Conference [Manitou-Stockholm, 2017] gave insight into his concern about the human environment and the need for global environmentalism. He opened his statement by pointing out that "We have made a global decision of immeasurable importance to which this meeting testifies: we have determined that we must control and harness the forces, which we have ourselves created."

Strong did not believe an environmental catastrophe was imminent, but he was concerned about the possibility. He pointed out that "Our whole work, our whole dedication is surely towards the idea of a durable and habitable planet." He was concerned that most people on the planet live in sustainable conditions. He asked, "can the great venture of human destiny be carried safely into a new century if our work is left in this condition? I, for one, do not believe it can."

According to Strong, the subject of the 1972 conference was the human environment, which he interpreted broadly. To Strong, "the human environment impinges upon the entire condition of man, and cannot be seen in isolation from war, and poverty, injustice and discrimination."

He said that "all nations must accept responsibility for the consequences of their own actions on environments outside their borders." He pointed out that "Our major motivation in gathering here is to consider recommendations, which can only be translated into action by international agreement. By far the major part of the burden of environmental management falls, however, upon national, governments operating as sovereign national states."

Strong believed in an incremental process to achieve his goals that was reminiscent of the Fabian Society. Strong said, "I cannot stress too strongly, the central importance of accepting this notion of ongoing process, of continuity, of adaptation, of steady evolution, in perception, in organization, in decision making and in action to protect and enhance the human environment."

Strong spoke as an ally of developing countries as well as a believer in growth. He said that "Many of the fundamental environmental problems of the developing countries derive from their very poverty and lack of resources and, in some cases, from inappropriate forms of development."

In addition, Strong believed in growth, but not just unlimited growth: "People must have access to more, not fewer opportunities to express their creative drives. But these can only be provided within a total system in which man's activities are in dynamic harmony with the natural order. To achieve this, we must control and redirect our processes of growth. We must rethink our concepts of the basic purposes of growth. We must see it in terms of enriching the lives and enlarging the opportunities of all mankind. And if this is so, it follows that it is the more wealthy societies — the privileged minority of mankind — which will have to make the most profound, even revolutionary changes in attitudes and values."

Strong argued that "The overall global goal of the United Nations environmental program must be to arrest the deterioration and begin the enhancement of the human environment." He concluded that "The basic task of this conference is to build in the minds of men the new vision of the larger, richer future which our collective will and energies can shape for all mankind, to build a program of concerted action which will make an important first step towards the realization for this vision; to build the new vehicle of international cooperation that will enable us to continue the long journey towards that creative and dynamic harmony between man and nature that will provide the optimum environment for human life on Planet Earth."

We see in Strong's opening statement to the 1972 Stockholm conference a recognition that change will be incremental and that wealthy countries will have to reconsider their use of natural resources.

The main outcome of the 1972 Stockholm Conference for the UN system was the establishment of four new entities that became known as the United Nations Environment Program (UNEP). The function of UNEP was to catalyze, coordinate, and stimulate action within the UN system. UNEP was not an authorization to execute or finance action.

Maurice Strong returned to Canada in 1976 when Prime Minister Pierre Trudeau asked him "to head the newly created national oil company, PetroCanada... He then became Chairman of the Canada Development Investment Corporation, the holding company for some of Canada's principal government-owned corporations" [Manitou-Strong, 2017]. Strong was a member of the Brundtland World Commission on Environment and Development presided over by Gro Harlem Brundtland. The Brundtland Commission summarized their work in the report *Our Common Future* [WCED, 1987].

A new world conference on the environment was held on the 20th anniversary of the 1972 Stockholm Conference. According to UNEP, "In June 1992 [Maurice Strong] was asked to lead another landmark meeting: the UN Conference on Environment and Development — best known as the Earth Summit — which was held in Rio de Janeiro, Brazil" [UNEP-Strong, 2018].

A major outcome of the United Nations Conference on Environment and Development (UNCED) was the creation of a program of action for achieving sustainable development into the 21st century called Agenda 21. Again, Strong's opening statement to the 1992 Rio Conference provided an update on Strong's vision for the future.

After welcoming conference attendees and dignitaries, Strong acknowledged that "despite significant progress made since 1972 in many areas, the hopes ignited at Stockholm remain largely unfulfilled [Manitou-Rio, 2017]." He told the assembly that "the environment, natural resources and life-support systems of our planet have continued to deteriorate, while global risks like those of climate change and ozone depletion have become more immediate and acute," based on the report "Our Common Future" by Gro Harlem Brundtland's World Commission on Environment and Development [WCED, 1987]. Strong went on to say "Yet all the environmental deterioration and risks we have experienced to date have occurred at levels of population and human activity that are much less than they will be in the period ahead."

Strong identified the central issues facing the conference as "patterns of production and consumption in the industrial world that are undermining the Earth's life-support systems; the explosive increase in population...deepening disparities between rich and poor that leave 75 per cent of humanity struggling to live; and an economic system that takes no account of ecological costs or damage."

Strong said that the "Population must be stabilized, and rapidly. If we do not do it, nature will, and much more brutally." In addition to population control, Strong pointed out that during the 20 years from the 1972 Stockholm Conference to the 1992 Rio Conference, "world GDP increased by $20 trillion. Yet 15 per cent of the increase accrued to developing countries," while over 70% went to already rich countries. "This is the other part of the population problem," he said, "the fact that every child born in the developed world consumes 20 to 30 times the resources of the planet than a third world child." He concluded that "The wasteful and destructive lifestyles of the rich cannot be maintained at the cost of the lives and livelihoods of the poor, and of nature."

He pointed out that he considered it "of the highest importance" that all governments represented at the Rio Conference translate the decisions made at the conference into policies and practices that implement the decisions. He specifically referred to Agenda 21.

Strong said Agenda 21 established "for the first time, a framework for the systemic, co-operative action required to effect the transition to sustainable development." He said "The issue of new and additional financial resources to enable developing countries to implement Agenda 21 is crucial and pervasive. This, more than any other issue, will clearly test the degree of political will and commitment of all countries to the fundamental purposes and goals of this Earth Summit." He said that "Agenda 21 measures for eradication of poverty and the economic enfranchisement of the poor provide the basis for a new worldwide war on poverty. Indeed, I urge you to adopt the eradication of poverty as a priority objective for the world community."

Strong pointed out that "countries and corporations which use energy and materials most efficiently are also those which are most successful economically...The transition to a more energy-efficient economy that weans us off our overdependence on fossil fuels is imperative to the achievement of sustainable development."

Strong concluded that "It is an exhilarating challenge to erase the barriers that have separated us in the past, to join in the global partnership that will enable us to survive in a more secure and hospitable world." He said that "The industrialized world cannot escape its primary responsibility to lead the way in establishing this partnership and making it work. Up to now, the damage inflicted on our planet has been done largely inadvertently. We now know what we are doing. We have lost our innocence. It would be more than irresponsible to continue down this path."

He ended his opening statement by telling the assembly "The road beyond Rio will be a long and difficult one; but it will also be a journey of renewed hope, of excitement, challenge and opportunity, leading as we move into the 21st century to the dawning of a new world in which the hopes and aspirations of all the world's children for a more secure and hospitable future can be fulfilled. This unprecedented responsibility is in your hands." The New York Times said that "donor nations agreed to provide $7 billion in aid to poorer ones, the sum was far short of the $70 billion that the United Nations said was needed annually" [Roberts, 2015].

One of the most significant actions taken at the 1992 Earth Summit in Rio de Janeiro was approval of Agenda 21, which was not a treaty, but a non-binding agreement that provided an action plan for achieving sustainable development. United States President George H.W. Bush attended the 1992 Earth Summit and signed the Agenda 21 agreement. Agenda 21 calls for countries with the most consumption to curb their consumption; countries with the largest populations to slow their population growth; and countries with the most wealth to increase financial and technological assistance to poorer countries to facilitate sustainable development worldwide. Figure 24-1 shows some key connections linking Maurice Strong to Agenda 21.

Strong helped launch the Earth Charter Initiative in 1994. The Earth Charter was supposed to be completed at the 1992 Earth Summit.

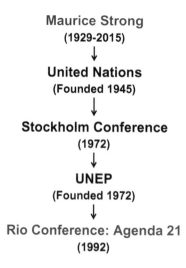

Figure 24-1. Connections from Maurice Strong to Agenda 21.

According to Strong, "the Earth Charter would set out the basic principles for the conduct of people and nations toward each other and the Earth to ensure our common future;" [Strong, 2000, page 202]. The Earth Charter is presented in Appendix A [Earth Charter, 2018]. Strong said the real goal of the Earth Charter was to "become like the Ten Commandments, like the Universal Declaration of Human Rights. It will become a symbol of the aspirations and the commitments of people everywhere." [Strong, 1998]

In its tribute to Maurice Strong upon his death in 2015, the Earth Charter Initiative said that "As a member of the Brundtland Commission and Secretary General of the Earth Summit in 1992, Mr. Strong took on the commitment to carry the idea of an Earth Charter forward (which was a recommendation made in the Brundtland Commission Report and according to him an unfinished piece of business of the Rio Earth Summit). Therefore, in 1994, together with Mikhail Gorbachev he launched the Earth Charter Initiative and became the co-Chair of the International Commission" [ECI-Strong, 2015].

Strong published the book *Where on Earth Are We Going?* in 2000 that presented his views at the time [Strong, 2000]. He began his book with a scenario that described what the world could be like in thirty years if the world did not take action to mitigate anthropogenic climate change. As of January 1, 2031, suggested Strong, changing weather patterns, global warming, environmental disasters, water shortages, the re-emergence of diseases that were once controlled, political turmoil, and disintegration of law and order could reduce the world's population by two thirds. His book was a call to action for leaders of government, business and environmental organizations to solve issues in the new millennium that transcended national boundaries. He said that "The time has come when we need to act both globally and locally, and that requires the cooperation of all of us, from individuals to grassroots groups to business, governments and supranational organizations." [Strong, 2000, page 5]

Strong was aware that he was considered part of a conspiracy to establish a world government. One of the possibilities he considered was transforming the United Nations into a supranational authority that could function as a world government. He tried to clarify his position by stating that a world government "is not necessary, not feasible and not desirable." [Strong, 2000, page 319] He pointed out that he was not implying that we should "aspire to a world without systems or rules." He was aware that "A chaotic world would pose equal or even greater danger. The challenge is to strike a balance so that the management of global affairs is responsive to the interests of all people in a secure and sustainable future. Such management must be guided by basic human values and make global organization conform to the reality of global diversity." [Strong, 2000, page 319]

Strong suggested that international organizations such as the United Nations could provide the basic elements "of an improved system of international agreements and international law and more streamlined international organizations to service and support the cooperation among governments and other key factors that will be required." [Strong, 2000, page 321] According to Strong, the rule of law was the key to the effective functioning of national and international societies. He believed that the "single greatest weakness of the existing international legal regime is the almost total lack of a capacity for enforcement." [Strong, 2000, page 342] In his view, "the codification, administration and enforcement of international law must become one of the principal functions of the United Nations in the period ahead." [Strong, 2000, page 341] This view of the United Nations contrasts with his previous statement that world government was not necessary, feasible, nor desirable.

Strong stepped aside as the UN envoy to North Korea in 2005 "after Tongsun Park, a South Korean with a scandalous past, was found to have been an unregistered lobbyist for Iraq in the United Nations oil-for-food program and to have invested $1 million in a company controlled by Mr. Strong. Mr. Strong was cleared of any involvement in the scandal" [Roberts, 2015]. Strong spent much of his time after 2005 in China as

Honorary Professor of Peking University in Beijing. Maurice Strong died in 2015.

25. International banking and globalization

Governments need access to revenue from the private sector in times of crisis, such as wars and natural catastrophes. Skousen pointed out that loaning "money to governments can be a very lucrative business" [Skousen, page 22]. In addition, loans to government from private funds give the lender influence that can be used to affect government decisions and the resolution of political issues [Quigley, 1966, pages 51–52].

Historically, financiers developed networks of banking centers and financial institutions to control lending to governments. Quigley called these financiers international bankers and pointed out that international bankers differ from conventional bankers in five ways [Quigley, 1966, page 52]: "(1) they were cosmopolitan and international; (2) they were close to governments and were particularly concerned with questions of government debts, including foreign government debts… (3) their interests were almost exclusively in bonds and very rarely in goods… (4) they were, accordingly, fanatical devotees of deflation…and of the gold standard (5) they were almost equally devoted to secrecy and the secret use of financial influence in political life."

Some international bankers were part of banking family dynasties such as the Rothschilds of Frankfurt and John Pierpont (J.P.) Morgan [Quigley, 1966, pages 51–52]. Morgan was a financier and banker. Banking family dynasties tended to be private firms until inheritance taxes made it necessary to preserve family wealth with "the immortality of corporate status for tax-avoidance purposes" [Quigley, 1966, page 52]

Quigley observed that the ability to extend and use credit helped expand the British Empire. The Bank of England, founded in 1694 by William Paterson, was an institution designed to centralize control of finance [Quigley, 1966, page 48]. A bank that made loans could create paper claims that exceeded its reserves of gold, thereby effectively creating money out of nothing [Quigley, 1966, page 49]. Furthermore, the bank received interest from credit, even though the reserves for the credit were only a fraction of the value of the credit. Reginald McKenna, Chancellor of the Exchequer in 1915–1916, told the stockholders of the bank he chaired in 1924 that "I am afraid the ordinary citizen will not like to be told that the banks can, and do, create money… And they who control the credit of the nation direct the policy of Governments and hold in the hollow of their hands the destiny of the people." [Quigley, 1966, page 325]

The United States economy was too dynamic in the beginning of the 20th century to be controlled by major banks. American colonists living under British rule before the American Revolution in the 18th century "were limited to using European coinage, barter and commodity money as their primary means of exchange" [FRBSF-History, 2018]. The colonists experienced shortages of foreign coins, and ineffective use of barter and commodity money. Many colonies chose to mint their coins and colonial banks issued paper currency backed by land or precious metals such as gold. Credit was provided by some merchants or other individuals.

The United States was formed after the War of Independence from Britain and the Unites States Constitution was ratified in 1789. Alexander Hamilton, serving as Secretary of the Treasury, proposed a federal banking system. Hamilton's plan was supported by commercial and financial interests in the more urbanized northeastern states. Thomas Jefferson, then serving as Secretary of State, represented agrarian interests and opposed Hamilton's central government plan. Jefferson preferred a more decentralized system of banks sanctioned by the states. Hamilton's plan won and the First Bank of the United States was chartered in 1791 and headquartered in Philadelphia.

An attempt to re-charter the First Bank in 1811 failed when Americans who were uncomfortable with a powerful central bank "dominated by big banking and money interests" succeeded in opposing the bank [FRE-History, 2018]. State banks emerged to fill the void. Many state banks began to issue paper currencies which had questionable value. Congress chartered the Second Bank of the United States in 1816 in an attempt to provide for the financial needs of the Federal government. The Second Bank operated until 1836 when renewal of its charter was vetoed by President Andrew Jackson who declared the existence and function of the Second Bank unconstitutional. As Thomas Jefferson had argued decades earlier, the Constitution did not "expressly authorize the federal government to charter a national bank or issue paper currency." [FRBSF-History, 2018]

The period between 1836 and the end of the American Civil War in 1865 was called the Free Banking Era. Nearly 8,000 banks chartered by the states were issuing their own paper notes. The National Banking Act was passed by Congress in 1863, during the Civil War, to provide reliable financing for the war effort. The Act "created a uniform national currency and permitted only nationally chartered banks to issue bank notes, but did not create a strong central banking structure." [FRBSF-History, 2018].

The weaknesses of the federal banking system created by the National Banking Act emerged when the industrial economy expanded after the Civil War. Many banks did not have enough cash on hand to meet demand when bank panics, or runs on banks, occurred. The periods of unusually high demand occurred when bank customers lost confidence in their bank, often after hearing about the failure of other banks. A banking panic in 1893 and another in 1907 required the intervention of J.P. Morgan, a financial mogul in the private sector. The panics led to a "growing consensus among all Americans that a central banking authority was needed to ensure a healthy banking system and to provide for an elastic currency." [FRE-History, 2018]

The Federal Reserve System was founded to help stabilize the United States banking system. The Federal Reserve Act was signed into law by President Woodrow Wilson in 1913. The original Federal

Reserve Board was selected by Wilson's adviser Edward M. House, who also recruited Walter Lippmann into the Wilson administration during World War I.

The Federal Reserve (also known as 'The Fed') has been designed to make monetary policy relatively independent of political pressure. Congress has mandated two policy goals for the Fed: maximize sustainable output and employment, and maintain stable prices. The Fed is expected to achieve these goals by "influencing the availability and the cost of money and credit to promote a healthy economy." [FRBSF-Monetary Policy, 2018]

The Federal Reserve System "has a two-part structure: a central authority called the Board of Governors in Washington, D.C., and a decentralized network of 12 Federal Reserve Banks located throughout the country. Monetary policy is set by the FOMC [Federal Open Market Committee], which includes members of the Board of Governors and presidents of the Reserve Banks." [FRBSF-Structure, 2018]

Quigley said the Fed has been controlled by international bankers since its inception. According to Quigley, international bankers in the 1920s were "determined to use the financial power of Britain and the United States to force all the major countries of the world to go on the gold standard and to operate it through central banks free from all political control, with all questions of international finance to be settled by agreements of such banks without interference from governments." [Quigley, 1966, page 326; Skousen, 1970, pages 21–24]

The goal of the international bankers, according to Quigley, was "… nothing less than to create a world system of financial control in private hands able to dominate the political system of each country and the economy of the world as a whole. This system was to be controlled in a feudalist fashion by the central banks of the world, acting in concert, by secret agreements arrived at in frequent private meetings and conferences." [Quigley, 1966, page 324]

26. Funding of globalization by the privileged minority

Money is needed to globalize the world. In addition to banks, many other organizations have been willing to provide the funds. Maurice Strong was concerned that the "privileged minority" [Strong, 2000, page 28] was not facing enough hardship to be motivated to support his call for action to address anthropogenic climate change. He did point out a few exceptions within the privileged minority that were willing to provide support through charitable and voluntary organizations. Strong said that "Particularly noteworthy are George Soros, who donates hundreds of millions of dollars a year through his Open Society Foundation, largely in the countries of the former Soviet Union, and the billionaire media genius Ted Turner, who made the largest single charitable contribution in U.S. history by committing $1 billion to support United Nations programs and activities. His generosity has since been topped by the computer software king Bill Gates, the first person ever to have his personal net worth reach $100 billion. They follow in the tradition of the great philanthropists of the past — notably the Rockefeller family, which continues in its current generation to set a remarkable example of enlightened and innovative philanthropic leadership." [Strong, 2000, page 28] Here we consider one source of funds, the Rockefeller family, to illustrate the loosely bound global network working for change.

26.1 *Amassing the Rockefeller fortune*

John Davison Rockefeller (1839–1937) was the source of the Rockefeller fortune [Tarbell, 1904; Yergin, 1992; EB-Standard Oil, 2018]. He started an oil refinery business in Cleveland, Ohio with Maurice B. Clark and Samuel Andrews in 1863. Rockefeller bought out Clark in 1865, the same year that the American Civil War ended. He spent the next few years building his business.

In 1867 Rockefeller invited Henry M. Flagler to become a partner in the firm of Rockefeller, Andrews, and Flagler. Their oil refinery business became the property of a new company called Standard Oil Company incorporated in Ohio in 1870 with headquarters in Cleveland, Ohio. In the intervening years, the company eliminated competitors, merged with other firms, and used its size and efficiency to extract favorable railroad rebates, effectively discounts, on railway freight rates to control over 90% of oil refining in the United States by 1880.

Rockefeller and associates combined affiliated companies in producing, refining, and marketing to form a holding company called Standard Oil Trust. The Standard Oil Trust Agreement was signed in 1882. According to Yergin [Yergin, 1992, page 43], "A board of trustees was set up, and in the hands of those trustees was placed the stock of all the entities controlled by Standard Oil. Shares in turn were issued in the trust; out of the 700,000 total shares, Rockefeller held 191,700 and Flagler, next, had 60,000." Yergin pointed out that "separate Standard Oil organizations were set up in each state to control the entities in those states." [Yergin, 1992, page 45] One of the organizations was Standard Oil of New Jersey. In 1885 the Standard Oil Trust moved its headquarters to New York.

The size and extent of the company raised concerns in Congress that the company was a monopoly. Ohio Senator John Sherman proposed an antitrust act that would authorize the Federal government to break up businesses that prohibit competition. The United States Congress passed the Sherman Antitrust Act in 1890.

The State of Ohio filed suit against Rockefeller and Standard Oil in 1892 under the antitrust act. The Ohio Supreme Court ordered the Trust to be dissolved in 1892, but the company appealed the decision and was able to continue operating through its New York headquarters.

In 1906, the Theodore Roosevelt Administration used the Sherman Antitrust Act to sue Standard Oil for conspiring to restrain trade [Yergin, 1992, page 108].

The U.S. Justice Department forced the company to breakup into many smaller companies in 1911. One of the spin-off companies took the name Standard Oil of New Jersey, and eventually became a part of Exxon and now ExxonMobil. The breakup of Standard Oil Trust made J.D. Rockefeller the richest man in the world [Yergin, The Prize, page 113]. Figure 26-1 highlights key events in the rise and fall of Standard Oil.

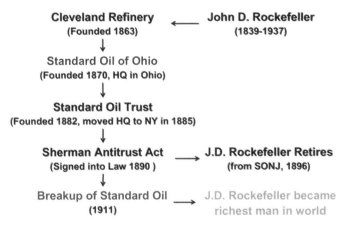

Figure 26-1. The Rise and Fall of Standard Oil.

26.2 *The Rockefeller Foundation*

John D. Rockefeller and his only son John D. Rockefeller, Jr. (1874–1960) founded the Rockefeller Foundation with Frederick Taylor Gates in New York State. The charter of the Rockefeller Foundation was accepted by the New York State Legislature on May 14, 1913. Gates was an oil and gas business and philanthropic advisor to John D. Rockefeller. Raymond B. Fosdick published a history of the early days of the Rockefeller Foundation [Fosdick, 1952]. His story covers the period from the time the charter of the Rockefeller Foundation was accepted by the New York State Legislature on May 14, 1913 until the end of

Fosdick's administration as President in 1948. Fosdick was very familiar with the Foundation's history after serving 12 years as President of the Foundation.

John D. Rockefeller and his son John D. Rockefeller, Jr. (1874–1960) founded the Rockefeller Foundation with Frederick Taylor Gates in 1913. Gates was a former Baptist minister who became an adviser in business and philanthropy to John D. Rockefeller. Figure 26-2 summarizes the founders of the Rockefeller foundation and selected Rockefeller family participants.

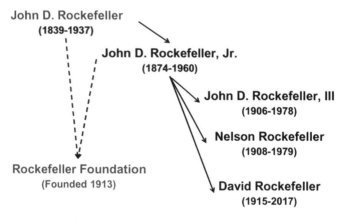

Figure 26-2.   Rockefeller Lineage and Founding of the Rockefeller Foundation.

Fosdick believed that Gates had the original idea to establish a Foundation in a letter he sent to John D. Rockefeller in 1905 [Fosdick, 1952, page 14]. In his letter, Fosdick said "It seems to me that either you and those who live now must determine what shall be the ultimate use of this vast fortune, or at the close of a few lives now in being it must simply pass into the unknown, like some other great fortunes, with unmeasured and perhaps sinister possibilities." [Fosdick, 1952, page 14] Following receipt of Gates' letter, John D., Sr. had many discussions with his son John D., Jr. and Gates about establishing a trust containing "considerable sums of money to be devoted to philanthropy, education, science, and religion." [Fosdick, 1952, page 15]

John D., Jr. earned his place in a three-cornered partnership between John D. Sr., John D. Jr., and Gates. John D., Jr. graduated from Brown University in 1897 and immediately went to work as an apprentice for his father. John D., Jr. later wrote "that if I was going to learn to help Father in the care of his affairs, the sooner my apprenticeship under his guidance began, the better." [Fosdick, 1952, page 3] Fosdick said of the relationship, "Few father-and-son relationships have been characterized by more genuine trust or deeper affection. For forty years they worked closely and intimately together." [Fosdick, 1952, page 3] The three-cornered partnership "was responsible for a group of foundations to which Mr. Rockefeller (John D., Sr.) contributed nearly half a billion dollars." [Fosdick, 1952, page 3]

Gates told John D., Sr. that "Your fortune is rolling up, rolling up like an avalanche! You must keep up with it! You must distribute it faster than it grows! If you do not, it will crush you and your children and your children's children!" [Fosdick, 1952, page 3] In addition to Gates' warning, Fosdick said that John D., Sr. was also influenced by an essay written by steel industrialist and philanthropist Andrew Carnegie (1835–1919) in 1889. Carnegie wrote that "the man who dies leaving behind him millions of available wealth, which was his to administer during life, will pass away 'unwept, unhonored, and unsung'... The man who dies thus rich dies disgraced." John D., Sr. concluded that "a man should make all he can and give all he can." [Fosdick, 1952, page 6]

John D., Sr. arrived at a set of principles to govern contributions in the 1880's: "His money should be given to work already organized and of proven worth; it should be work of a continuing character which would not disappear when his gifts were withdrawn; the contributions, where possible, should be made on conditional terms so as to stimulate contributions by others; and finally ... his money should make for strength rather than weakness and should develop in the beneficiary a spirit of independence and self-reliance." [Fosdick, 1952, page 6]

In 1909, John D., Sr. signed a deed of trust that turned over approximately $50,000,000 dollars in shares of Standard Oil Company of New Jersey stock to three trustees: his son John D., Jr.; his son-in-law Harold

McCormick; and Frederick T. Gates. The trust was called "The Rockefeller Foundation," and one of the first orders of business for the trustees was to obtain a corporate charter.

A bill was introduced in the United States Senate to incorporate The Rockefeller Foundation in March 1910. According to Fosdick, John D., Sr. was seeking government control to make sure that the funds could not be used improperly and, if they were, the people's representatives in Congress had the power to protect his wish that the fund always be used for the public welfare.

Congress kept the bill dangling until 1913 because its members were concerned that John D., Sr. was attempting to enact a scheme that could perpetuate his vast wealth. When Congress adjourned in 1913 without passing the bill, John D., Sr.'s advisers asked the New York State Legislature in Albany to incorporate The Rockefeller Foundation. The Legislature acted within months of the adjournment of Congress and The Rockefeller Foundation was incorporated in 1913. The mission of the Foundation was "to promote the well-being of mankind throughout the world." [Fosdick, 1952, page 20]

Four decades later, in the dawn of the Cold War following World War II, Fosdick said that humanity had to reconcile itself to the "grim necessities which today's problem of security brings to all of us" [Fosdick, 1952, page 288]. He wrote that "in a deep and ultimate sense, it is still one world, one human race, one common destiny. That was the high faith that lay behind the creation of the Foundation, and on that faith the future must depend." [Fosdick, 1952, page 288]

Today, the work of the Rockefeller Foundation [RF-Work, 2018] is "to secure the fundamentals of human well-being around the world." The website explains that "For more than 100 years, The Rockefeller Foundation has brought people together around the globe to try to solve the world's most challenging problems and promote the well-being of humanity. Today, in a world capable of so much, it is unacceptable that there are still so many with so little. That's why the Rockefeller Foundation fights to secure the fundamentals of human well-being — health, food, energy, jobs — so they're within reach for everyone, everywhere in the world. Our approach is grounded in what we've seen work over more than a century: It's inspired by science, rigorous about data, brings together and empowers others, and is focused on real results that improve people's lives."

26.3 *International banker David Rockefeller*

Maurice Strong mentioned that he first met David Rockefeller, son of John D. Rockefeller, Jr., while Strong was a young man trying to get a job with the United Nations in New York. Over the course of their careers Strong and Rockefeller were members of some of the same organizations or worked on some of the same projects. A few examples of international organizations that link Maurice Strong of the United Nations with banker David Rockefeller are illustrated in Figure 26-3. David's oldest brother John Davison Rockefeller, III is shown because of the role he played leading the Rockefeller Foundation. We show below how the connections

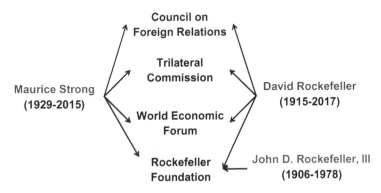

Figure 26-3. A Few Connections between Maurice Strong and Members of the Rockefeller Family.

occurred by reviewing selected aspects of David's biography and learning more about his concept of internationalism.

David Rockefeller was born in 1915. He was the son of John D. Rockefeller, Jr. and Abigail Green Aldrich Rockefeller (1874–1948). Abigail was the sister of Winthrop Aldrich (1885–1974). Winthrop was a lawyer who served as Chairman of Chase Bank following a stock selling scandal in 1933. He supported two structural banking reforms that were passed by Congress as a result of the scandal: the Glass-Stegal Act, which separated commercial banking from investment banking, and the Security Act which authorized the Securities and Exchange Commission (SEC). The Security Act required corporations to register their stock and report significant financial disclosures [Rockefeller, 2002, page 125].

David attended Harvard during the Great Depression which began with the Wall Street Collapse in 1929. He graduated from Harvard in 1936 and stayed for another year studying economics at the graduate level. While at Harvard he had a chance to learn from economists engaged in a debate over government's role in economics. British economist John Maynard Keynes (1883–1946) was advocating for government intervention to stimulate the economy. Economists that supported the free market were concerned that Keynesian government intervention would replace the market with permanent government control of the economy.

David left Harvard to attend the London School of Economics (LSE) for a year. He observed that "In those days the LSE was widely considered a hot bed of socialism and radicalism. Founded by the Webbs in the 1890s to help achieve their Fabian Socialist goal of a just society based on a more equal distribution of wealth, its walls had always given shelter to men and women who tested the limits of orthodoxy." [Rockefeller, 2002, page 82]. We saw in our discussion of Fabianism that Fabian Society members Sidney and Beatrice Webb founded LSE in 1895.

David said that political science professor Harold Laski contributed significantly to the reputation of the LSE during the 1920s and

1930s. Laski "enthralled well-filled classrooms with his eloquent Marxist lectures." [Rockefeller, 2002, page 82] Referring to Laski, David "found the intellectual content of his lectures superficial and often devious and deceptive. They seemed more propaganda than pedagogy." [Rockefeller, 2002, page 82]

David was more favorably impressed with the more conservative economics department at the LSE. David was tutored by Austrian economist Friedrich von Hayek who won the Nobel Prize in Economics in 1974 for his work on money, the business cycle, and capital theory in the 1930s and 1940s. Like other economists at the LSE, Hayek trusted the market and opposed Keynesian interventionist economics. Hayek wrote *The Road to Serfdom* between 1940 and 1943 as a warning that government control of economic decision-making by central planning could lead to tyranny in the form of fascism or national socialism [Hayek, 1944]. At the time the book was written, Nazi Germany and Fascist Italy were in control of much of Europe and threatening the British Isles.

David left the LSE in 1938 so that he could finish his graduate work at the University of Chicago. Economists at the University of Chicago tended to be advocates of the free market and believed the market and natural pricing mechanisms were more likely to sustain economic growth than government intervention.

David returned to New York in 1939 to write his doctoral dissertation in economics. He completed his dissertation and received his doctorate from the University of Chicago in 1940. His education helped him realize that he was "a pragmatist who recognizes the need for sound fiscal and monetary policies to achieve optimum economic growth." [Rockefeller, 2002, page 92] Furthermore, he recognized that affordable social safety nets are needed in society.

David worked with his brothers to advance their interests in philanthropy. The Rockefeller philanthropic tradition "required that we be generous with our financial resources and involve ourselves actively in the affairs of our community and the nation." [Rockefeller, 2002, page 145] The Rockefeller brothers founded the Rockefeller Brothers Fund (RBF) in 1940. The brothers believed that, over time, the RBF would give them the opportunity "to work together and to forge a philanthropic philosophy that reflected our generation's values and objectives." [Rockefeller, 2002, page 141]

After David completed his dissertation, he worked for New York Mayor Fiorello La Guardia (1882–1947) until the United States entered World War II in 1941. David's mother, Abigail, impressed upon him that he had a 'duty' to join the war effort.

David enlisted as a private in the Army in 1942. He received corporal's stripes when he finished basic training and was assigned to Counter-Intelligence. In 1943 he applied to and was accepted into Officer Candidate School (OCS). He was commissioned a second Lieutenant when he completed OCS and was assigned to Military Intelligence. He was sent to Algiers and charged with developing an intelligence network. His service during the war taught him the value of networking: "I discovered the value of building contacts with well-placed individuals as a means of achieving concrete objectives." [Rockefeller, 2002, page 122]

David took a position in Chase National Bank in 1946 after the war. Even though his uncle Winthrop Aldrich was Chairman of the bank, David did not start at the top. David "began as an assistant manager, the lowest officer rank, in the foreign department at an annual salary of $3,500." [Rockefeller, 2002, page 137]

David saw that the bank had little international presence and believed he could make a difference by expanding the bank's presence internationally. He traveled to many regions of the world, including Europe, to see where he could make the most difference. He realized that Latin America and the Caribbean were underserved markets. During this time David met young Maurice Strong in New York in the late 1940s.

By the early 1950s, the Chase branch system in the Caribbean "had emerged as the most dynamic part of our overseas operations." [Rockefeller, 2002, page 133] David was eager to apply the Caribbean strategy of introducing branch banks, buying into local banks, and expanding into new lending activity to expand into other parts of the world. David was on track to become Chair of Chase in 1969.

In the meantime, the first 12 years of the Rockefeller Brothers Fund (RBF), which was founded in 1940, did not have an endowment. The RBF was funded by individual annual donations from the brothers. By the 1950s, their "individual annual contributions to the RBF had grown to the point that we were able to support organizations which individual brothers had initiated or in which one of us had a special interest." [Rockefeller, 2002, page 140]

David Rockefeller's international travels in the 1930s and his experience overseas in World War II sharpened his awareness of international affairs, which he "would develop through active involvement with the Council on Foreign Relations, the Carnegie Endowment for International Peace, and International House of New York." [Rockefeller, 2002, page 145] In addition, Chase created an International Advisory Committee in the 1960s to strengthen the bank's access to leaders around the world. When he retired in 1981, David became chairman of the IAC. [Rockefeller, 2002, pages 208–209]

He was also a member of the Bilderberg group and became chair of the Council on Foreign Relations in 1985. David wrote that "Bilderberg is really an intensely interesting annual discussion group that debates issues of significance to both Europeans and North Americans — without reaching consensus." [Rockefeller, 2002, page 411] By comparison, we saw in our discussion of Cecil Rhodes that the Council on Foreign Relations emerged from Rhodes Secret Society and the Round Table (see Figure 20-1).

David founded The Trilateral Commission in 1973 with Zbigniew Brzezinski and Jimmy Carter. The purpose of the Commission was to engage Japan, an economic power, with North America and Europe to discuss problems that he believed would require trilateral efforts by the three regions [Trilateral Commission Tribute, 2017] One of the first reports of the Club of Rome was *Limits to Growth* [Meadows, et al., 1972]. The report presented the results of a study that found the exponential growth of the human population given a finite supply of resources could not support the expanding human ecosystem. According to Rockefeller, Brzezinski called the report a "pessimist manifesto" [Rockefeller, 1991]. Energy scholar Vaclav Smil said *Limits to Growth* was "easily the most widely publicized, and hence the most influential, forecast of the 1970s, if not the last one-third of the twentieth century" [Smil, 2003, page 168]. Smil analyzed the study and concluded that the "report pretended to capture the intricate interactions of population, economy, natural resources, industrial production, and environmental pollution with less than 150 lines of simple equations using dubious assumptions to tie together sweeping categories of meaningless variables." [Smil, 2003, page 169]

The Trilateral Commission published a study in 1991 [MacNeill *et al.*, 1991], prior to the 1992 Earth Summit Conference in Rio de Janeiro, that showed an interconnection between the world's economy and the earth's ecology. David Rockefeller, as North American Chairman of the Trilateral Commission, wrote the Foreward for the book, and Maurice Strong, as Secretary General of the United Nations Conference on Environment and Development, wrote the Introduction for the book.

In 2004, Brzezinski said that trilateralism was initially the strategy of seeking "trilateral cooperation between the three major democratic centers of economic and political power — between North America, Western Europe, and Japan". In Brzezinski's view, trilateralism "was the key to global stability and progress; that therefore trans-Atlantic cooperation had to be wedded to trans-Pacific cooperation and that in the world at large democracy plus prosperity equal influence and power and that such power and influence should be harnessed for the common good, both for the sake of self-interest and of the moral imperative." [Brzezinski, 2004]

The turbulence of the Vietnam War years and President Richard Nixon's resignation in 1973 after the Watergate break-in made the members of David Rockefeller's family reconsider the common desires that bound them together as a family. They found that they shared a set of common desires: "to create a more just world that was free of racial intolerance and bigotry; to eliminate poverty; to improve education; and to figure out how the human race could survive without destroying the environment." [Rockefeller, 2002, page 322]

David realized that the conflict in his immediate family actually was present in his extended family. David's brother John D. Rockefeller, III had spent his life in philanthropy. Normally soft-spoken, John resisted the efforts of his hard-charging younger brother Nelson to take control of the Rockefeller philanthropic organizations. Nelson was a nationally prominent Republican politician who would serve as Vice President under Gerald R. Ford (1913–2006) after Nixon's resignation.

John had served as chairman of the Rockefeller Foundation since the 1950s, and "viewed himself as the legitimate 'heir' of the Rockefeller tradition of philanthropy, which he also considered the core value of the Rockefeller family and the only activity that could over time hold family members together." [Rockefeller, 2002, page 338] David wrote that "Philanthropy was John's field, and he resented Nelson's assertions that it was he, rather than his older brother, who should guide the future of the family's primary philanthropies, particularly RBF. [Rockefeller, 2002, 339]

David observed that part of the conflict was due to a leftward shift in John's political views. John had spent quite a bit of time interviewing young people in the early 1970s in an attempt to understand their point of view. In 1973, John published his findings in *The Second American Revolution* [Rockefeller, J.D. III, 1973].

John D. Rockefeller, III expressed his belief that the turmoil in the streets of the United States in the 1960s and early 1970s meant that the United States was at a turning point in history. His thesis was that "instead of being overwhelmed by our problems, we must have faith that they can be resolved, that we can achieve a society in which humanistic values predominate." [Rockefeller, J.D. III, 1973, page xiv]

John identified "two very broad sources of revolutionary change in our society. One is people, individually and in groups, who are concerned about justice and freedom and receiving a fair share of the fruits of our society. The other is impersonal and materialistic, stemming from economic growth, new knowledge and technological innovations, international rivalries." [Rockefeller, J.D. III, 1973, page 5]

John referred to the revolution as a humanistic revolution because it emerged from the first source of change: "the wants and needs and aspirations of people." [Rockefeller, J.D. III, 1973, page 6] It was a revolution because it could lead to far-reaching social change, and possibly the replacement of one government by another. The humanistic revolution would "harness the forces of economic and technological change in the service of humanistic values." [Rockefeller, J.D. III, 1973, page 6]

John believed the central meaning of the youth movement was "to achieve a person-centered society, instead of one built around materialism and large impersonal institutions which breed conformity rather than individuality and creativity." [Rockefeller, J.D. III, 1973, page 28] He said that the older generation should help sustain the youth movement rather than suppress it. He argued that the youth movement had several positive accomplishments: for example, the youth movement had stimulated social awareness of issues, youth were more concerned about ideas and spiritualism than materialism, and that the causes of poverty are "to be found less in the failings of the poverty-stricken than in social imbalances and discrimination." [Rockefeller, J.D. III, 1973, pages 31–32]

John was aware that the "steady growth of centralized government must ultimately result in a statist and bureaucratic pattern quite unlike anything we have known in the past. The only alternative is for government and other sectors of society to collaborate in leading us back to a balanced system, based on the bedrock of individual initiative." [Rockefeller, J.D. III, 1973, page 108] He believed that the Federal government should facilitate participation of the private sector and other levels of government in helping solve social problems.

David Rockefeller summarized John's 1973 book by writing that John believed "all wisdom reposed in the young and that the older generation, which had made such a mess of the world, should look to their children for guidance." [Rockefeller, 2002, page 339] According to David, John's experience had "strengthened his instinctive sympathies for the underdog and the underclass." [Rockefeller, 2002, page 339] David's brother Nelson did not want John, with his apparent shift in political outlook, to fund organizations that would oppose Nelson's political aspirations.

David published an autobiography entitled *Memoirs* [Rockefeller, 2002]. He said that "For more than a century ideological extremists at either end of the political spectrum have seized upon well-publicized incidents such as my encounter with Castro to attack the Rockefeller family for the inordinate influence they claim we wield over American political and economic institutions. Some even believe we are part of a secret cabal working against the best interests of the United States, characterizing my family and me as 'internationalists' and of conspiring with others around the world to build a more integrated global political and economic structure — one world, if you will. If that's the charge, I stand guilty, and I am proud of it." [Rockefeller, 2002, page 405]

David believed that "the free flow of investment capital, goods, and people across borders will remain the fundamental factor in world economic growth and in the strengthening of democratic institutions everywhere." [Rockefeller, 2002, 406] Therefore, the United States should not be isolationist but should embrace its global responsibilities. He declared "we must all be internationalists." [Rockefeller, 2002, page 406]

David Rockefeller died in 2017.

27. Picking up the environmental baton: Barack Obama

Someone needed to fill the void left when Maurice Strong stepped away from the United Nations in 2005. Former Vice President Al Gore would have been in a powerful position to take over leadership of the global environmental movement had he won the 2000 election for President of the United States. Gore's loss to George W. Bush forced Gore to operate in the private sector.

Gore developed and presented a slide show to the general public warning of the dangers of global warming. In 2006 he wrote and starred in the documentary film "An Inconvenient Truth" which documented his efforts to educate the public about climate change. The film won two Academy Awards, one for Best Documentary Feature and one for Best Original Song, from the American Academy of Motion Picture Arts and Sciences in 2007 [Gore-Oscars, 2007]. In addition, Gore shared the 2007 Nobel Peace Prize with the Intergovernmental Panel on Climate Change (IPCC) "for their efforts to build up and disseminate greater knowledge about man-made climate change, and to lay the foundations for the measures that are needed to counteract such change" [Gore-Nobel Prize, 2007]. The IPCC is a panel that operates under the auspices of the United Nations and was established in 1988 by the World Meteorological Organization and the United Nations Environment Programme (UNEP). It is worth recalling that Maurice Strong helped found UNEP in 1972 and was its first Executive Director.

A new global environmental leader did not emerge until 2008 when then Democrat Senator Barack Obama won the United States Presidential election over his Republican opponent Senator John McCain. According to Gore, he hoped to answer the question of how the world would solve the challenges posed by anthropogenic climate change by saying that "The turning point came in 2009…with the inauguration of a new President [Obama] in the United States, who immediately shifted priorities to focus on building the foundation for a new low-carbon economy." [Gore, 2009, page 394]

Obama was inaugurated in January 2009 and promptly selected several environmentalists to manage his departments. Some notable members of the Obama Administration are shown in Figure 27-1.

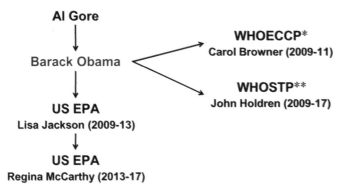

Figure 27-1. Some Climate Policy Players in the Obama Administration.
*White House Office of Energy and Climate Change Policy.
**White House Office of Science and Technology Policy.

Lawyer and environmentalist Carol Martha Browner served as Obama's Director of the White House Office of Energy and Climate Change Policy from 2009 to 2011. The office was eliminated as a standing White House Office in 2011 and its climate and energy work was taken over by the Domestic Policy Council.

John P. Holdren was senior advisor to President Barack Obama on science and technology issues. Holdren had degrees in aerospace engineering and theoretical plasma physics. He was Assistant to the President for Science and Technology, Director of the White House Office of Science and Technology Policy from 2009 to 2017, and Co-Chair of the President's Council of Advisors on Science and Technology (PCAST).

Lisa Jackson, a chemical engineer, served as the Administrator of the United States Environmental Protection Agency (EPA) from 2009 to 2013. Regina McCarthy took over Jackson's duties at the EPA in Obama's second term (2013 to 2017). She was an environmental health and air quality expert.

The Obama Administration implemented many climate and energy policies during the period from 2009 to 2017, which corresponds to Obama's two Presidential terms. The Obama White House Archives (OWHA) pointed out that President Obama had made "a historic commitment to protecting the environment and addressing the impacts of climate change" [OWHA Climate, 2018]. According to the OWHA document, "President Obama believes that no challenge poses a greater threat to our children, our planet, and future generations than climate change — and that no other country on Earth is better equipped to lead the world towards a solution. That's why under President Obama's leadership, the United States has done more to combat climate change than ever before, while growing the economy."

The Obama Administration sought to protect the environment and address possible effects of climate change in six areas: cut carbon pollution; expand the clean energy economy; lead global efforts on climate change; protect our climate, air and water; cut energy waste; and protect natural resources. Some of these efforts are summarized in the tables below.

Table 27-1 presents some of the steps taken by the Obama Administration to improve the environment by cutting carbon pollution.

Table 27-1. Steps to Cut Carbon Pollution.

A. A plan called the Clean Power Plan was announced in August 2015. The plan set the first national standards for regulating carbon pollution from power plants [US EPA Clean Power Plan, 2018].
B. Fuel economy standards were set so that fuel economy would be increased in passenger vehicles, and medium and heavy-duty trucks.
C. New energy efficiency standards for appliances and equipment were set.
D. A new strategy to reduce methane missions from such sources as oil and gas development and landfills were developed and implemented.
E. Reduced greenhouse gas emissions from Federal government facilities and equipment.

Table 27-2. Steps to Expand the Clean Energy Economy.

A. Passed the American Recovery and Reinvestment Act in 2009. The Recovery Act was the largest single investment in clean energy in history. It provided "more than $90 billion in strategic clean energy investments and tax incentives to promote job creation and the deployment of low-carbon technologies, and leveraging approximately $150 billion in private and other non-federal capital for clean energy investments" [OWHA Recovery Act, 2018]
B. Approved the first-ever large-scale renewable energy projects on Federal lands.
C. Launched the Clean Energy Investment Initiative in an attempt to stimulate $2 billion of expanded private sector investment in solutions to climate change [OWHA Clean Energy Investment, 2018].
D. Launched the Clean Energy Savings for All Initiative "to ensure that every household has options to choose to go solar and put in place additional measures to promote energy efficiency" [OWHA Clean Energy Savings, 2018]. The Initiative was particularly focused on expanding access for low- and moderate-income communities and creating a more inclusive workforce [OWHA Climate, 2018].
E. Expanded and modernized the electric grid by upgrading technology and reforming the permit process.
F. Introduced a 21st Century Clean Transportation Plan to fund the infrastructure for fueling electric vehicles.
G. "Dedicated new federal resources for economic diversification, job creation, training, and other employment services for workers and communities impacted by layoffs at coal mines and coal-fired power plants." [OWHA Climate, 2018]

Table 27-2 presents some of the steps taken by the Obama Administration to expand the clean energy economy by scaling up renewable energy. Step G in the list was needed to address a publicly perceived war on coal that arose in a November 7, 2008 interview with the San Francisco Chronicle. Then presidential candidate Obama said "If somebody wants to build a coal powered plant, they can, it's just that it will bankrupt them because they are going to be charged a huge sum for all that greenhouse gas that's being emitted."

Table 27-3 presents some of the steps taken by the Obama Administration to lead global efforts on climate change.

Table 27-3. Steps to Lead Global Efforts on Climate Change.

A. The highlight of this effort was to support the Paris Climate Agreement at COP21.
B. Forged a historic joint agreement to reduce greenhouse gas emissions with China.
C. Pledged $3 billion to support the Green Climate Fund (GCF). The purpose of funding the GCF was "to reduce carbon pollution and strengthen resilience in developing countries, especially the poorest and most vulnerable" [OWHA Climate, 2018].
D. Sought a commitment by member states of the Organization for Economic Cooperation and Development (OECD) to "dramatically reduce financing for coal-fired power plants overseas" [OWHA Climate, 2018].

Table 27-4 presents some of the steps taken by the Obama Administration to protect our climate, air and water.

Table 27-4. Steps to Protect the Climate, Air and Water.

A. Set national limits on power plant and other industrial emissions of mercury and other air pollutants.
B. Set new standards for cleaner gasoline and vehicles.
C. Reformed the Toxic Substances Control Act (TSCA) to give the EPA additional authority to help protect Americans from chemicals that could adversely affect health.
D. Wrote the Clean Water Rule to help clarify which waters are covered by the Clean Water Act, which is designed to prevent pollution of covered waters.
E. Overhauled the national offshore energy program following the Deepwater Horizon disaster.

Table 27-5 presents some of the steps taken by the Obama Administration to cut energy waste by improving energy efficiency.

Table 27-5. Steps to Cut Energy Waste.

A. Sought to improve energy efficiency in facilities operated by utilities, manufacturers, businesses, school districts, cities and states using the Better Buildings Challenge.
B. New energy efficiency standards were set for appliances and equipment.
C. Directed federal agencies to develop adaptation plans to "reduce the vulnerability of federal programs, assets, and investments to the impacts of climate change" [OWHA Climate, 2018].
D. Launched the Climate Education and Literacy Initiative to educate the public about the climate change challenge.
E. Sought to improve national climate change preparedness by developing a Climate Resilience Toolkit (CRT) and launching the Climate Data Initiative (CDI). The CRT provides information to help communities prepare for the impact of climate change. The CDI was designed to expand the availability of freely available climate data to stimulate innovation and private sector entrepreneurship.

Table 27-6 presents some of the steps taken by the Obama Administration to protect natural resources.

Table 27-6. Steps to Protect Natural Resources.

A. Signed the Omnibus Public Land Management Act of 2009. The Act "designated more than 2 million acres of Federal wilderness and protected thousands of miles of trails and more than one thousand miles of rivers" [OWHA Climate, 2018].
B. Created the first United States marine monument in the Atlantic Ocean off the New England coast. The monument preserved sea canyons and underwater mountains.
C. Permanently protected more than 550 million acres of America's public lands and waters.

President Barack Obama applied the resources of the United States Federal government to the challenges he believed were the consequence of anthropogenic climate change. However, many of the steps outlined in Tables 27-1 through 27-6 were not enacted as laws, but as regulations. Agreements such as the Paris Climate Agreement were signed by the

Executive, but did not have the authority of a Treaty because they were not approved by the United States Senate. Once inaugurated in January 2017, President Donald Trump and his administration began removing regulations and agreements enacted during the Obama Administration. For example, the Trump Administration withdrew the United States from the COP21 Paris Climate Agreement in June 2017 because the Trump Administration was concerned that the agreement would have a greater negative impact on the United States economy than a positive impact on the global environment. The difference in climate and energy policies between the Obama Administration and the Trump Administration illustrates the point made in Section 14 that the Liberal-Progressive perspective on energy and the Conservative perspective on energy are competing energy visions.

Did President Obama live up to the hope expressed by Al Gore in 2009 [Gore, 2009, page 394] that Obama would solve the challenges posed by anthropogenic climate change? The election of Republican Donald Trump in 2016 as presidential successor to Democrat Barack Obama suggests that Obama's tenure as President of the United States was not enough to satisfy Gore's hope.

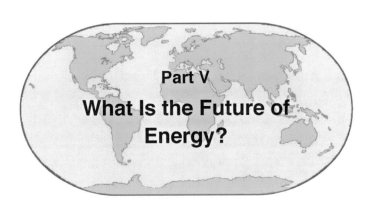

28. Can the obstacles be overcome?

Many global trends influence energy policy. Some of the trends were led by people such as Cecil Rhodes, U Thant, Maurice Strong, John D. Rockefeller, and Barack Obama.

- Cecil Rhodes was an elitist (some say racist) and imperialist who established scholarships to attempt to continue his policies beyond the grave using acolytes from future generations.
- United Nations Secretary General U Thant hoped the environmental crisis would be the challenge the world needed to motivate the establishment of a new world order and lead to a more civilized and generous way of life for all of the world's people.
- Maurice Strong was an industrialist and self-professed environmental sinner who became an environmentalist and high level official of the United Nations. He was associated with several Non-Governmental Organizations (NGOs) that allowed him to connect with the world of finance. In addition, he was associated with banks and foundations throughout his career. In 1995, he was appointed senior adviser to the president of World Bank [Rosset and Russell, 2007]. The modern goals of the World Bank are "to end extreme poverty and promote shared prosperity in a sustainable way" [World Bank, 2018].
- John D. Rockefeller amassed considerable wealth using questionable business practices and then established a trust that protected his family's wealth for generations. His family's foundations helped fund environmental policies that can influence government policy on a global scale. Their goal is to guide the world away from fossil fuels and toward a renewable energy future.
- President Barack Obama used resources of the United States Federal government to help the world face the challenges he associated with anthropogenic climate change.

Many of the trends considered in Part IV are leading to the same destination: a supranational authority controlled by a ruling class. The United Nations is the current contender for becoming the supranational authority. Agenda 21, the Earth Charter, and the COP21 Paris Climate Agreement have established a roadmap to a sustainable energy mix in a low carbon economy. The timetable for this roadmap is urgent; the energy transition needs to occur before atmospheric temperature increases more than 2°C above pre-industrial levels. Depending on the temperature change forecast you choose, the transition to a low carbon economy might have to be completed in three or four decades, despite observations made by analysts such as Vaclav Smil and Daniel Yergin that the energy transition might not be complete until the end of the 21st century.

28.1 *Competing economic visions*

A common thread running through many global trends is the view that the leaders of the trends tend to be members of the ruling class seeking to centralize control of the global economy. By contrast, opponents of the trend toward collectivism are people that prefer individualism. The resulting clash of economic perspectives continues a clash that began in the early 20th century when Keynes's view of economics competed with Hayek's view. Hayek and Keynes were proponents of two competing visions of economics: Hayek was a proponent of free market capitalism, while Keynes advocated government intervention in the economy using monetary and fiscal policies. Hayek realized that the central issue was whether or not government could successfully plan an economy. Here we consider two competing visions of a state economy: free market capitalism, and government controlled central planning.

Free market capitalism depends on competition in the marketplace. Competition manifests itself as a division of the control of the means of production among many people acting independently. Competition depends on the Rule of Law whereby government is bound by well-known rules that apply equally to everyone. The Rule of Law safeguards equality in legal proceedings and is the legal embodiment of freedom in a competitive society. The power of one person over another is minimized by freedom of choice. If one person is unwilling to satisfy our wishes, we can find someone else who will. The result is a free market that depends on competition to set prices. Entrepreneurs use the free market to determine what actions they should take to satisfy consumer demand.

The alternative to free market capitalism is government controlled central planning. Central planning on a large scale requires power to coerce. The state establishes rules, such as laws and regulations, to exercise control over the individual. Police or the military are typically used by the state to enforce its rules.

Central planners can plan for competition, but competition reduces central control. This is not a call for anarchy, but recognition of the limits of government. The role of the state includes establishing an economic environment that supports competition, breaks up monopolies, and prevents fraud and deception. The state must have totalitarian power to run a directed economy. In a modern state, staffs of experts are needed to direct complex systems of interrelated activities. The ultimate authority in a totalitarian state resides in a commander-in-chief or dictator.

The introduction of a degree of competition reduces the level of control imposed by the state. Central planning supersedes planning by the individual and reduces the choices available to the individual. Consequently, as the extent of central planning increases, the freedom of the individual is diminished. Democracy is an obstacle to centralized control of economic activity because democratic assemblies cannot function as effective planning agencies. Hence, centralized planning can lead to dictatorship.

28.2 *Global feudalism*

Proponents of anthropogenic climate change believe that global measures and a sense of urgency are needed to meet environmental challenges. It is worth recalling from Section 12.3 that United States Secretary of State John Kerry pointed out the importance of global cooperation while attending the 2015 COP21 Paris Climate meeting: "If all the industrial nations went down to zero emissions ... it wouldn't be enough, not when more than 65 percent of the world's carbon pollution comes from the developing world."

Globalists tend to favor open borders and multinational agreements such as the European Union Charter of Fundamental Rights, the North American Free Trade Agreement (NAFTA), the Trans-Pacific Partnership (TPP), Agenda 21 and the Earth Charter. The British exit from the European Union (BREXIT) and the election of Donald Trump as President of the United States show voter concern on both sides of the Atlantic about the value of multinational agreements and the loss of national sovereignty, including loss of control of foreign policy and immigration policy.

Supranational agreements designed to meet the challenges associated with anthropogenic climate change are favored by advocates of central planning on a global scale. In *The Road to Serfdom*, Hayek [Hayek, 1944] warned that tyranny could result from government controlled central planning. From this perspective, we could see the emergence of a global ruling class that rejects fossil fuels in favor of an urgent transition to renewable energy. The growth of power by a global centralized authority will reduce individual freedom to make choices and could lead to the emergence of global feudalism: a global ruling class and everyone else.

The Goldilocks Policy minimizes the risk of global feudalism by establishing a framework that can be adopted worldwide. The free market can be used to determine which energy sources are most economic at any point in time during the transition to a sustainable energy mix.

29. Selecting our energy future

Our future energy mix depends on choices we make, which depends, in turn, on energy policy. Several criteria need to be considered when establishing energy policy. We need to consider the capacity of the energy mix, its cost, safety, reliability, and effect on the environment. We need to know that the energy mix has the capacity to meet our needs and have the reliability to be available when it is needed. The energy mix should have a negligible or positive effect on the environment, and it should be safe. When we consider cost, we need to consider both tangible and intangible costs associated with each component of the energy mix.

Historical trends in energy consumption suggest that society is continuing a trend toward decarbonization, or the reduction in the relative amount of carbon in combustible fuels. The next step in decarbonization would be a transition to natural gas. A greater reliance on natural gas rather than coal or oil would reduce the emission of greenhouse gases into the atmosphere.

The 21st century energy mix will depend on technological advances, including some advances that cannot be anticipated, and on choices made by society. There are competing visions for reaching a sustainable energy mix. Some people see an urgent need to replace fossil fuels with sustainable/renewable energy sources because human activity is driving climate change. Others believe that it is necessary to replace fossil fuels with sustainable/renewable energy sources, but the need is not urgent. They argue that the economic health of society outweighs possible climate effects. If the energy transition is too fast, it could significantly damage the global economy. If the energy transition is too slow, damage to the environment could be irreversible.

The "Goldilocks Policy for Energy Transition" is designed to establish a middle ground between these competing visions. We need the duration of the energy transition to be just right; that is, we need to adopt a reasonable plan of action that reduces uncertainty with predictable public policy and reduces environmental impact.

Future energy demand is expected to grow substantially as global population increases and developing nations seek a higher quality of life. Social concern about nuclear waste and proliferation of nuclear weapons is a significant deterrent to reliance on nuclear fission power. These concerns are alleviated to some extent by the safety record of the modern nuclear power industry. Society's inability to resolve the issues associated with nuclear fusion make fusion an unlikely contributor to the energy mix until at least the middle of the 21st century. If nuclear fusion power is allowed to develop and eventually becomes commercially viable, it could become the primary energy source. Until then, the feasible sources of energy for use in the future energy mix are fossil fuels, nuclear fission, and renewable energy.

Based on historical data, we could plan an energy transition to a sustainable energy mix by the latter half of the 21st century. The European Union is operating on this timetable by hoping to complete the EU Supergrid by 2050. In addition, natural gas can serve as a transition fuel because it is relatively abundant, continues the trend to decarbonization, halves greenhouse gas emissions relative to oil and coal combustion, and requires reasonably affordable infrastructure changes that take advantage of available technology. A natural gas infrastructure would be a step toward a hydrogen economy infrastructure in the event that hydrogen becomes a viable energy carrier. The development of a game-changing technology, such as commercial nuclear fusion, would substantially accelerate the transition to a sustainable energy mix.

Appendix A
The Goldilocks Story

A widely known version of the Goldilocks story describes the adventures of a girl named Goldilocks who found herself in the cottage of a family of three bears: papa bear, mama bear and baby bear. Goldilocks was attracted to the cottage by the smell of freshly made porridge. The porridge was so hot that the bears had decided to go for a walk so that the porridge could cool. While the bears were on their walk, Goldilocks entered the cottage and found three bowls of porridge on a table. She tasted the porridge and found one that was too hot, one that was too cold, and one that was just right. After eating the porridge that was just right she entered another room and found three chairs. She sat on the chairs and found that one was too hard, one was too soft, and one was just right. She rested on the chair that was just right. When she felt sleepy, Goldilocks went upstairs to a bedroom with three beds. She found one that was too high, one that was too low, and one that was just right. The three bears eventually came home and discovered that someone had been in the house. They looked for the intruder and found Goldilocks asleep upstairs in the bed that was just right. Goldilocks awoke, saw the bears looking at her, and ran away.

Appendix B
The Earth Charter [Earth Charter, 2018]

Preamble

We stand at a critical moment in Earth's history, a time when humanity must choose its future. As the world becomes increasingly interdependent and fragile, the future at once holds great peril and great promise. To move forward we must recognize that in the midst of a magnificent diversity of cultures and life forms we are one human family and one Earth community with a common destiny. We must join together to bring forth a sustainable global society founded on respect for nature, universal human rights, economic justice, and a culture of peace. Towards this end, it is imperative that we, the peoples of Earth, declare our responsibility to one another, to the greater community of life, and to future generations.

Earth, our home

Humanity is part of a vast evolving universe. Earth, our home, is alive with a unique community of life. The forces of nature make existence a demanding and uncertain adventure, but Earth has provided the conditions essential to life's evolution. The resilience of the community of life and the well-being of humanity depend upon preserving a healthy biosphere with all its ecological systems, a rich variety of plants and animals, fertile soils, pure waters, and clean air. The global environment with its finite resources is a common concern of all peoples. The protection of Earth's vitality, diversity, and beauty is a sacred trust.

The global situation

The dominant patterns of production and consumption are causing environmental devastation, the depletion of resources, and a massive extinction of species. Communities are being undermined. The benefits of development are not shared equitably and the gap between rich and poor is widening. Injustice, poverty, ignorance, and violent conflict are widespread and the cause of great suffering. An unprecedented rise in human population has overburdened ecological and social systems. The foundations of global security are threatened. These trends are perilous — but not inevitable.

The challenges ahead

The choice is ours: form a global partnership to care for Earth and one another or risk the destruction of ourselves and the diversity of life. Fundamental changes are needed in our values, institutions, and ways of living. We must realize that when basic needs have been met, human development is primarily about being more, not having more. We have the knowledge and technology to provide for all and to reduce our impacts on the environment. The emergence of a global civil society is creating new opportunities to build a democratic and humane world. Our environmental, economic, political, social, and spiritual challenges are interconnected, and together we can forge inclusive solutions.

Universal responsibility

To realize these aspirations, we must decide to live with a sense of universal responsibility, identifying ourselves with the whole Earth community as well as our local communities. We are at once citizens of different nations and of one world in which the local and global are linked. Everyone shares responsibility for the present and future well-being of the human family and the larger living world. The spirit of human solidarity and kinship with all life is strengthened when we live with reverence for the mystery of being, gratitude for the gift of life, and humility regarding the human place in nature.

We urgently need a shared vision of basic values to provide an ethical foundation for the emerging world community. Therefore, together in hope we affirm the following interdependent principles for a sustainable way of life as a common standard by which the conduct of all individuals, organizations, businesses, governments, and transnational institutions is to be guided and assessed.

PRINCIPLES

I. RESPECT AND CARE FOR THE COMMUNITY OF LIFE

1. Respect Earth and life in all its diversity.
   a. Recognize that all beings are interdependent and every form of life has value regardless of its worth to human beings.
   b. Affirm faith in the inherent dignity of all human beings and in the intellectual, artistic, ethical, and spiritual potential of humanity.

2. Care for the community of life with understanding, compassion, and love.
   a. Accept that with the right to own, manage, and use natural resources comes the duty to prevent environmental harm and to protect the rights of people.
   b. Affirm that with increased freedom, knowledge, and power comes increased responsibility to promote the common good.

3. **Build democratic societies that are just, participatory, sustainable, and peaceful.**
    a. Ensure that communities at all levels guarantee human rights and fundamental freedoms and provide everyone an opportunity to realize his or her full potential.
    b. Promote social and economic justice, enabling all to achieve a secure and meaningful livelihood that is ecologically responsible.
4. **Secure Earth's bounty and beauty for present and future generations.**
    a. Recognize that the freedom of action of each generation is qualified by the needs of future generations.
    b. Transmit to future generations values, traditions, and institutions that support the long-term flourishing of Earth's human and ecological communities. In order to fulfill these four broad commitments, it is necessary to:

II. ECOLOGICAL INTEGRITY

5. **Protect and restore the integrity of Earth's ecological systems, with special concern for biological diversity and the natural processes that sustain life.**
    a. Adopt at all levels sustainable development plans and regulations that make environmental conservation and rehabilitation integral to all development initiatives.
    b. Establish and safeguard viable nature and biosphere reserves, including wild lands and marine areas, to protect Earth's life support systems, maintain biodiversity, and preserve our natural heritage.
    c. Promote the recovery of endangered species and ecosystems.
    d. Control and eradicate non-native or genetically modified organisms harmful to native species and the environment, and prevent introduction of such harmful organisms.

e. Manage the use of renewable resources such as water, soil, forest products, and marine life in ways that do not exceed rates of regeneration and that protect the health of ecosystems.
  f. Manage the extraction and use of non-renewable resources such as minerals and fossil fuels in ways that minimize depletion and cause no serious environmental damage.

6. **Prevent harm as the best method of environmental protection and, when knowledge is limited, apply a precautionary approach.**
  a. Take action to avoid the possibility of serious or irreversible environmental harm even when scientific knowledge is incomplete or inconclusive.
  b. Place the burden of proof on those who argue that a proposed activity will not cause significant harm, and make the responsible parties liable for environmental harm.
  c. Ensure that decision making addresses the cumulative, long-term, indirect, long distance, and global consequences of human activities.
  d. Prevent pollution of any part of the environment and allow no build-up of radioactive, toxic, or other hazardous substances.
  e. Avoid military activities damaging to the environment.

7. **Adopt patterns of production, consumption, and reproduction that safeguard Earth's regenerative capacities, human rights, and community well-being.**
  a. Reduce, reuse, and recycle the materials used in production and consumption systems, and ensure that residual waste can be assimilated by ecological systems.
  b. Act with restraint and efficiency when using energy, and rely increasingly on renewable energy sources such as solar and wind.
  c. Promote the development, adoption, and equitable transfer of environmentally sound technologies.

d. Internalize the full environmental and social costs of goods and services in the selling price, and enable consumers to identify products that meet the highest social and environmental standards.
e. Ensure universal access to health care that fosters reproductive health and responsible reproduction.
f. Adopt lifestyles that emphasize the quality of life and material sufficiency in a finite world.

8. **Advance the study of ecological sustainability and promote the open exchange and wide application of the knowledge acquired.**
   a. Support international scientific and technical cooperation on sustainability, with special attention to the needs of developing nations.
   b. Recognize and preserve the traditional knowledge and spiritual wisdom in all cultures that contribute to environmental protection and human well-being.
   c. Ensure that information of vital importance to human health and environmental protection, including genetic information, remains available in the public domain.

III. SOCIAL AND ECONOMIC JUSTICE

9. **Eradicate poverty as an ethical, social, and environmental imperative.**
   a. Guarantee the right to potable water, clean air, food security, uncontaminated soil, shelter, and safe sanitation, allocating the national and international resources required.
   b. Empower every human being with the education and resources to secure a sustainable livelihood, and provide social security and safety nets for those who are unable to support themselves.
   c. Recognize the ignored, protect the vulnerable, serve those who suffer, and enable them to develop their capacities and to pursue their aspirations.

10. **Ensure that economic activities and institutions at all levels promote human development in an equitable and sustainable manner.**
    a. Promote the equitable distribution of wealth within nations and among nations.
    b. Enhance the intellectual, financial, technical, and social resources of developing nations, and relieve them of onerous international debt.
    c. Ensure that all trade supports sustainable resource use, environmental protection, and progressive labor standards.
    d. Require multinational corporations and international financial organizations to act transparently in the public good, and hold them accountable for the consequences of their activities.

11. **Affirm gender equality and equity as prerequisites to sustainable development and ensure universal access to education, health care, and economic opportunity.**
    a. Secure the human rights of women and girls and end all violence against them.
    b. Promote the active participation of women in all aspects of economic, political, civil, social, and cultural life as full and equal partners, decision makers, leaders, and beneficiaries.
    c. Strengthen families and ensure the safety and loving nurture of all family members.

12. **Uphold the right of all, without discrimination, to a natural and social environment supportive of human dignity, bodily health, and spiritual well-being, with special attention to the rights of indigenous peoples and minorities.**
    a. Eliminate discrimination in all its forms, such as that based on race, color, sex, sexual orientation, religion, language, and national, ethnic or social origin.
    b. Affirm the right of indigenous peoples to their spirituality, knowledge, lands and resources and to their related practice of sustainable livelihoods.

c. Honor and support the young people of our communities, enabling them to fulfill their essential role in creating sustainable societies.

d. Protect and restore outstanding places of cultural and spiritual significance.

IV. DEMOCRACY, NONVIOLENCE, AND PEACE

13. Strengthen democratic institutions at all levels, and provide transparency and accountability in governance, inclusive participation in decision making, and access to justice.

a. Uphold the right of everyone to receive clear and timely information on environmental matters and all development plans and activities which are likely to affect them or in which they have an interest.

b. Support local, regional and global civil society, and promote the meaningful participation of all interested individuals and organizations in decision making.

c. Protect the rights to freedom of opinion, expression, peaceful assembly, association, and dissent.

d. Institute effective and efficient access to administrative and independent judicial procedures, including remedies and redress for environmental harm and the threat of such harm.

e. Eliminate corruption in all public and private institutions.

f. Strengthen local communities, enabling them to care for their environments, and assign environmental responsibilities to the levels of government where they can be carried out most effectively.

14. Integrate into formal education and life-long learning the knowledge, values, and skills needed for a sustainable way of life.

a. Provide all, especially children and youth, with educational opportunities that empower them to contribute actively to sustainable development.

b. Promote the contribution of the arts and humanities as well as the sciences in sustainability education.
c. Enhance the role of the mass media in raising awareness of ecological and social challenges.
d. Recognize the importance of moral and spiritual education for sustainable living.

15. Treat all living beings with respect and consideration.
a. Prevent cruelty to animals kept in human societies and protect them from suffering.
b. Protect wild animals from methods of hunting, trapping, and fishing that cause extreme, prolonged, or avoidable suffering.
c. Avoid or eliminate to the full extent possible the taking or destruction of non-targeted species.

16. Promote a culture of tolerance, nonviolence, and peace.
a. Encourage and support mutual understanding, solidarity, and cooperation among all peoples and within and among nations.
b. Implement comprehensive strategies to prevent violent conflict and use collaborative problem solving to manage and resolve environmental conflicts and other disputes.
c. Demilitarize national security systems to the level of a non-provocative defense posture, and convert military resources to peaceful purposes, including ecological restoration.
d. Eliminate nuclear, biological, and toxic weapons and other weapons of mass destruction.
e. Ensure that the use of orbital and outer space supports environmental protection and peace.
f. Recognize that peace is the wholeness created by right relationships with oneself, other persons, other cultures, other life, Earth, and the larger whole of which all are a part.

THE WAY FORWARD

As never before in history, common destiny beckons us to seek a new beginning. Such renewal is the promise of these Earth Charter principles. To fulfill this promise, we must commit ourselves to adopt and promote the values and objectives of the Charter. This requires a change of mind and heart. It requires a new sense of global interdependence and universal responsibility. We must imaginatively develop and apply the vision of a sustainable way of life locally, nationally, regionally, and globally. Our cultural diversity is a precious heritage and different cultures will find their own distinctive ways to realize the vision. We must deepen and expand the global dialogue that generated the Earth Charter, for we have much to learn from the ongoing collaborative search for truth and wisdom. Life often involves tensions between important values. This can mean difficult choices. However, we must find ways to harmonize diversity with unity, the exercise of freedom with the common good, short-term objectives with long-term goals. Every individual, family, organization, and community has a vital role to play. The arts, sciences, religions, educational institutions, media, businesses, nongovernmental organizations, and governments are all called to offer creative leadership. The partnership of government, civil society, and business is essential for effective governance. In order to build a sustainable global community, the nations of the world must renew their commitment to the United Nations, fulfill their obligations under existing international agreements, and support the implementation of Earth Charter principles with an international legally binding instrument on environment and development. Let ours be a time remembered for the awakening of a new reverence for life, the firm resolve to achieve sustainability, the quickening of the struggle for justice and peace, and the joyful celebration of life.

ORIGIN OF THE EARTH CHARTER

The Earth Charter was created by the independent Earth Charter Commission, which was convened as a follow-up to the 1992 Earth Summit in order to produce a global consensus statement of values and principles for a sustainable future. The document was developed over nearly a decade through an extensive process of international consultation, to which over five thousand people contributed. The Charter has been formally endorsed by thousands of organizations, including UNESCO and the IUCN (World Conservation Union). For more information, please visit www.EarthCharter.org

References

Allen, G. and L. Abraham (1971): *None Dare Call It Conspiracy*, Cutchogue, New York: Buccaneer Books.

Archibald, D. (2018): 11,000 years of Climate Swings in the Northern Hemisphere, http://climate.geologist-1011.net/ accessed 1-27-2018.

Bailey, R. (1997): "Who is Maurice Strong?" *The National Review*, September 1, 1997.

Bakewell, Charles M. (1901): "A Democratic Philosopher and His Work. Thomas Davidson: Born Oct. 25, 1840. Died Sept. 14, 1900," *International Journal of Ethics*, Volume 11, July 1901.

Bocking, S. (2012): "Nature on the Home Front: British Ecologists' Advocacy for Science and Conservation," *Environment and History*, June issue, DOI: 10.3 197/096734012X13303670112894; accessed July 31, 2017 at https://www.researchgate.net/profile/Stephen_Bocking/publication/233510180_Nature_on_the_Home_Front_British_Ecologists%27_Advocacy_for_Science_and_Conservation/links/568ab87008aebccc4e1a1819/Nature-on-the-Home-Front-British-Ecologists-Advocacy-for-Science-and-Conservation.pdf.

Boden, T.A., G. Marland, and R.J. Andres (2015): "National CO2 Emissions from Fossil-Fuel Burning, Cement Manufacture, and Gas Flaring: 1751–2011," Carbon Dioxide Information Analysis Center, Oak Ridge National Laboratory, U.S. Department of Energy, doi: 10.3334/CDIAC; data reported at the U.S. EPA website, http://www3.epa.gov/climatechange/ghgemissions/global.html, accessed December 30, 2015.

Brooks, A.C. (2010): *The Battle*, New York: Basic Books.

Burnham, J. (1941): *The Managerial Revolution*, first published by New York: John Day Company; reprinted in 1972 by Westport, Connecticut: Greenwood Press.

Carnegie, A. (1889): "The Gospel of Wealth," first published by The North American Review in 1889, republished in 2017 by New York: Carnegie Corporation of New York; available online at https://www.carnegie.org/publications/the-gospel-of-wealth/, accessed March 31, 2018.

Carson, R. (1962): *Silent Spring*, New York: Houghton Mifflin.

Chaffetz, J. (2018): *The Deep State*, Broadside Books, HarperCollins.

Chatham House (2018): "History," Chatham House website https://www.chathamhouse.org/about/history accessed March 6, 2018.

Chevalier, J-M (2009): *The New Energy Crisis: Climate, Economics and Geopolitics*, New York: Palgrave MacMillan, St. Martin's Press.

Clark, K. (1964): *John Ruskin — Selected Writings*, London: Penquin Books printed in 1991; first published as *Ruskin Today* by New York: Holt, Rinehart and Winston in 1964.

COP21 Agreement (2015): "Climate Action — Paris Agreement," European Commission website, last update was December 23, 2015, http://ec.europa.eu/clima/policies/international/negotiations/future/index_en.htm, accessed January 5, 2016.

Duffy, M. (1974): "Who's Who: Walter Lippmann," accessed March 14, 2018 at http://www.firstworldwar.com/bio/lippmann.htm.

Earth Charter (2018): "The Earth Charter," http://earthcharter.org/discover/download-the-charter/; accessed March 31, 2018.

EB — A.L. Strong (2018): "Anna Louise Strong, American Journalist and Scholar," Editors of the Encyclopedia Britannica, website https://www.britannica.com/biography/Anna-Louise-Strong, last update March 22, 2018; accessed March 25, 2018.

EB — Standard Oil (2018): "Standard Oil Company and Trust," Editors of the Encyclopedia Britannica, website https://www.britannica.com/topic/Standard-Oil-Company-and-Trust, last update January 25, 2018; accessed March 23, 2018.

ECI — Strong (2015): "Tribute to Maurice F. Strong (1929–2015)," Earth Charter Initiative, published online November 30, 2015, http://earthcharter.org/news-post/tribute-to-maurice-f-strong-1929-2015/, accessed March 24, 2018.

Ehrlich, P.R. (1968): *The Population Bomb*, New York: Ballentine Books.

Ehrlich, P.R. and A.H. Ehrlich (2009): "The Population Bomb Revisited," *The Electronic Journal of Sustainable Development,* Volume 1 Number 3.

Ehrlich, P.R. A.H. Ehrlich, and J.P. Holdren (1973): *Human Ecology: Problems and Solutions*, San Francisco: W.H. Freeman and Company.

Ehrlich, P.R. A.H. Ehrlich, and J.P. Holdren (1977): *Ecoscience: Population, Resources, Environment*, San Francisco: W.H. Freeman and Company.

Eisenhower, D. (1961): "Farewell Address to the Nation" on January 17, 1961; https://www.eisenhower.archives.gov/all_about_ike/speeches/farewell_address.pdf, accessed March 9, 2018.

Engels, F. (1886): *Ludwig Feuerbach and the End of Classical German Philosophy*, first published in *Die Neue Zeit* (1886); https://www.marxists.org/archive/marx/works/download/Marx_Ludwig_Feurbach_and_the_End_of_German_Classical_Philosop.pdf accessed July 26, 2017.

Fabian Society (2017a): "The Fabian Story," Fabian Society website, http://www.fabians.org.uk/about/the-fabian-story/ accessed October 4, 2017.

Fabian Society (2017b): "About the Fabian Society," Fabian Society website, http://www.fabians.org.uk/about/ accessed October 4, 2017 accessed October 4, 2017.

Fanchi, J.R. and C.J. Fanchi (2017): *Energy in the 21st Century*, Singapore: World Scientific.

FRBSF-History (2018): "What is the Fed: History," by Federal Reserve Bank of San Francisco, https://www.frbsf.org/education/teacher-resources/what-is-the-fed/history/, accessed March 27, 2018.

FRBSF-Monetary Policy (2018): "What is the Fed: Monetary Policy," by Federal Reserve Bank of San Francisco, https://www.frbsf.org/education/teacher-resources/what-is-the-fed/monetary-policy/, accessed March 27, 2018.

FRBSF-Structure (2018): "What is the Fed: Structure," by Federal Reserve Bank of San Francisco, https://www.frbsf.org/education/teacher-resources/what-is-the-fed/structure/, accessed March 27, 2018.

FRE-History (2018): "History of the Federal Reserve," by the Federal Reserve, https://www.federalreserveeducation.org/about-the-fed/history/2006andbeyond, accessed March 27, 2018.

Firestone, B.J. (2001): *The United Nations under U Thant, 1961–1971*, London: The Scarecrow Press, 2001.

Flint, J. (1974): *Cecil Rhodes*, Boston: Little, Brown and Company.

Fosdick, R.B. (1952): *The story of the Rockefeller Foundation*, New York: Harper.

Foster, J.B. (2000): *Marx's ecology: Materialism and nature*, New York: Monthly Review Press.

Foster, J.B. (2000): "E. Ray Lankester, Ecological Materialist: An Introduction to Lankester's 'The Effacement of Nature by Man,'" *Organization & Environment*, Volume 13, Number 2, June, pages 233–235.

Foster, J.B. (2002): "Marx's ecology in historical perspective," International Socialism, website https://mronline.org/johnbellamyfoster/wp-content/uploads/sites/9/2014/10/Foster-2002-Marxs_Ecology_in_Historical_Perspective.pdf accessed July 31, 2017.

Founders (1776): Declaration of Independence, U.S. Department of the Interior, accessed July 28, 2017 at http://www.constitution.org/us_doi.pdf.

Friedman, T. (2014): "From Putin, A Blessing In Disguise," Op-Ed, New York Times, March 19.

Goldberg, J. (2014, June 20): "Of the Bureaucrats, by the Bureaucrats, for the Bureaucrats," National Review Online, accessed June 25, 2014.

Gore, A. (2009): *Our Choice: A Plan to Solve the Climate Crisis*, New York: Rodale-Melchir Media.

Gore-Nobel Prize (2007): "The Nobel Peace Prize 2007," https://www.nobelprize.org/nobel_prizes/peace/laureates/2007/, accessed April 3, 2018.

Gore-Oscars (2007): The 79th Academy Awards, https://www.oscars.org/oscars/ceremonies/2007, accessed April 3, 2018.

Hayek, F. (1944): *The Road to Serfdom*, London: Routledge.

Hegel, G.W.F. (1820): *Philosophy of Right*, translated by S.W. Dyde, printed in Kitchener, Ontario: Batoche Books, 2001. https://www.history.com/this-day-in-history/first-issue-of-the-new-republic-published.

History.com Staff (Nov. 7, 2007): "This day in history: November 7, 1914, First Issue of The New Republic Published," accessed March 14, 2018.

Hofmeister, J. (2010): *Why We Hate the Oil Companies*, New York: Palgrave MacMillan, St. Martin's Press.

Holland, H.D. (2006): "The oxygenation of the atmosphere and oceans," Phil. Trans. R. Soc. B, Vol. 361, 903–915.

Hourdin, F., T. Mauritsen, A. Gettelman, J-C. Golaz, V. Balaji, Q. Duan, D. Folini, D. Ji, D. Klocke, Y. Qian, F. Rauser, C. Rio, L. Tomassini, M. Watanabe, and D. Williamson (2017): "The Art and Science of Climate Model Tuning," *Bulletin of the American Meteorological Society*, March 2017, 589–602

IPCC (2001): Climate Change 2001, The Scientific Basis, United Nations Intergovernmental Panel on Climate Change (IPCC), https://www.ipcc.ch/ipccreports/tar/wg1/pdf/WGI_TAR_full_report.pdf, accessed February 20, 2018.

IPCC (2007): Climate Change 2007 report, United Nations Intergovernmental Panel on Climate Change (IPCC), http://ipcc.ch/publications_and_data/publications_and_data_reports.shtml#1, accessed December 31, 2015.

IPCC (2018): "Organization," Intergovernmental Panel on Climate Change (IPCC) website http://www.ipcc.ch/organization/organization.shtml accessed April 15, 2018.

Keenan, T.F., I.C. Prentice, J.G. Canadell, C.A. Williams, H. Wang, M. Raupach, and G.J. Collatz(2016): "Recent pause in the growth rate of atmospheric CO_2 due to enhanced terrestrial carbon uptake," *Nature Communications*, Volume 7, Article number 13428; DOI: 10.1038/ncomms13428; www.nature.com/naturecommunications accessed October 22, 2018.

Lankester, E.R. (1913): "The Effacement of Nature by Man," Chapter XXVIII in *Science from an Easy Chair; A Second Series*, New York: H. Holt and Company.

Lenin, V.I. (1901): "The Abolition of the Antithesis Between Town and Country. Particular Questions Raised by the 'Critics,'" first published as Chapter IV in *The Agrarian Question and the 'Critics of Marx'* by Zarya, December 1901; *Lenin Collected Works*, Foreign Languages Publishing House, 1961, Moscow, Volume 5, pages 103–222; accessed September 6, 2017 at https://www.marxists.org/archive/lenin/works/1901/agrarian/index.htm.

Lenin CEC (1918): "The Fundamental Law of Land Socialization," Decree of the Central Executive Committee, February 19, 1918; accessed September 6, 2017 at http://www.barnsdle.demon.co.uk/russ/land.html.

Lewis, J. (1985): "Birth of the EPA," EPA Journal, accessed September 18, 2017 at https://archive.epa.gov/epa/aboutepa/birth-epa.html.

Liebig, J. (1840): *Organic Chemistry in its Applications to Agriculture and Physiology*, edited by L. Playfair; London: Printed for Taylor and Walton, Booksellers and Printers to the University; accessed July 28, 2017 at https://archive.org/details/organicchemistry00liebrich.

Lippmann, W. (1919): *The Political Scene; An Essay on the Victory of 1918*, New York: H. Holt and Company, accessed March 20, 2018 at https://ia801406.us.archive.org/25/items/politicalscenea01lippgoog/politicalscenea01lippgoog.pdf.

Locke, J. (1680): *Two Treatises of Government*, Norwalk, Connecticut: The Easton Press Books that Changed the World, 1991.

Locke, J. (1690): "Concerning Human Understanding," the Pennsylvania State University, Electronic Classics Series, Jim Manis, Faculty Editor, Hazleton, PA 18201-1291, published 1999; accessed July 27, 2017 at ftp://ftp.dca.fee.unicamp.br/pub/docs/ia005/humanund.pdf.

Lofgren, M. (2014): "Essay: Anatomy of the Deep State," http://billmoyers.com/2014/02/21/anatomy-of-the-deep-state/ accessed March 9, 2018.

MacNeill, J., P. Winsemius, and T. Yakushiji (1991): *Beyond Interdependence*, A Trilateral Commission Book.

Madison, J. (1792): "Property," from *The Founders' Constitution*, edited by P.B. Kurland and R. Lerner (Chicago: University of Chicago Press, 1987), Volume 1, pages 598–599.

Malthus, T.R. (1798): *An Essay on the Principle of Population As It Affects the Future Improvement of Society, with Remarks on the Speculations of Mr. Goodwin, M. Condorcet and Other Writers* (1st edition). London: J. Johnson in St Paul's Church-yard.

Manitou — Rio (2017): "Maurice Strong's Opening statement to the Rio Summit (3 June 1992)," Manitou Foundation, https://www.mauricestrong.net/index.php/opening-statement6, accessed March 24, 2018.

Manitou — Stockholm (2017): "United Nations Conference on the Human Environment, 1972; Opening Statement by Maurice Strong, Secretary-General of the Conference," Manitou Foundation, https://www.mauricestrong.net/index.php/speeches-remarks3/103-stockholm, accessed March 24, 2018.

Manitou — Strong (2017): "Maurice Strong: Short Biography," Manitou Foundation, https://www.mauricestrong.net/index.php/short-biography-mainmenu-6, accessed March 21, 2018.

Mann, M.E., R.S. Bradley, and M.K. Hughes (1998): "Global-Scale Temperature Patterns and Climate Forcing over the Past Six Centuries," *Nature 392*, pages 779–787.

Mann, M.E., R.S. Bradley, and M.K. Hughes (1999): "Northern Hemisphere Temperatures during the Past Millennium: Inferences, Uncertainties, and Limitations," *Geophysical Research Letters 26*, pages 759–762.

Mann, M.E., R.S. Bradley, M.K. Hughes (2004): "Corrigendum: Global-Scale Temperature Patterns and Climate Forcing over the Past Six Centuries," *Nature 430*, page 105.

Märald, E. (2002): "Everything Circulates: Agricultural Chemistry and Recycling Theories in the Second Half of the Nineteenth Century," *Environment and History*, Volume 8, pages 65–84.

Marx, K. (1841): "The Difference between the Democritean and Epicurean Philosophy of Nature," first published in 1902; obtained from *Marx-Engels Collected Works*, Volume 1 by Progress Publishers; accessed July 26, 2017 at https://murdercube.com/files/Philosophy/Marx,_Karl_-_Doctoral_Thesis_-_The_Difference_Between_the_Democritean_and_Epicurean_Philosophy_of_Nature.pdf.

Marx, K. (1845): "Theses on Feurbach," accessed September 8, 2017 at http://www.marx2mao.com/M&E/TF45.html.

Marx, K. (1866): February 13 letter from Marx to Engels in *Marx & Engels Collected Works*, Volume 42, Letters 1864–68, copyright 2010 by Lawrence & Wishart, Electric Book, ISBN 978-1-84327-986-0; accessed July 29, 2017 at http://www.hekmatist.com/Marx%20Engles/Marx%20&%20Engels%20Collected%20Works%20Volume%2042_%20Ka%20-%20Karl%20Marx.pdf.

Marx, K. (1887): *Capital: A Critique of Political Economy, Volume I*, First English edition of 1887 including Fourth German edition changes; accessed July 29, 2017 at https://www.marxists.org/archive/marx/works/download/pdf/Capital-Volume-I.pdf.

Marx, K. (1894): *Capital: A Critique of Political Economy, Volume III*, edited by Friedrick Engels and completed by him 11 years after Marx's death; accessed July 29, 2017 at https://www.marxists.org/archive/marx/works/download/pdf/Capital-Volume-III.pdf.

Marx and Engels (1848): "Manifesto of the Communist Party," in *Communist Manifesto and Other Writings*, translated by Martin Milligan, Norwalk, Connecticut: The Easton Press Collector's Edition, 2005.

Masson-Delmotte, V., M. Schulz, A. Abe-Ouchi, J. Beer, A. Ganopolski, J.F. González Rouco, E. Jansen, K. Lambeck, J. Luterbacher, T. Naish, T. Osborn, B. Otto-Bliesner, T. Quinn, R. Ramesh, M. Rojas, X. Shao, and A. Timmermann (2013): "Information from Paleoclimate Archives," In Climate Change 2013: The Physical Science Basis. Contribution of Working Group I to the Fifth Assessment Report of the Intergovernmental Panel on Climate Change [Stocker, T.F., D. Qin, G.-K. Plattner, M. Tignor, S.K. Allen, J. Boschung, A. Nauels, Y. Xia, V. Bex, and P.M. Midgley (eds.)]. Cambridge University Press, Cambridge, United Kingdom and New York, NY, USA.

McIntyre, S. and R. McKitrick (2003): "Corrections to the Mann, *et al.* (1998) Proxy Data Base and Northern Hemisphere Average Temperature Series," *Environment and Energy 14* (6), pages 751–771.

McKitrick, R. (2005): "What is the Hockey Stick Debate About?" APEC Study Group, Australia, April 4, 2005; see also https://www.uoguelph.ca/~rmckitri/research/APEC-hockey.pdf, accessed February 20, 2018.

Meadows, D.H., D.L. Meadows, J. Randers, and W.W. Behrens III (1972): *Limits to Growth*, Washington, D.C.: Potomac Associates.

Mische, P.M. and M.A. Ribeiro (1998): "Ecological Security and the United Nations System," Chapter 10 in *The Future of the United Nations: Potential for the Twenty-first Century*, edited by C.F. Alger, New York: United Nations University Press.

NWE-Rhodes (2018): "Cecil Rhodes," New World Encyclopedia, http://www.newworldencyclopedia.org/entry/Cecil_Rhodes#cite_note-9, accessed February 20, 2018.

OGCI Founding (2018): "Oil and Gas Climate Initiative," http://oilandgasclimateinitiative.com/, accessed September 28, 2018.

OGCI US (2018): http://oilandgasclimateinitiative.com/oil-and-gas-climate-initiative-welcomes-chevron-exxonmobil-and-occidental-petroleum-into-its-international-membership/ accessed September 21, 2018.

Orwell, G. (1937): *The Road to Wigan Pier*, London: Victor Gollancz Ltd.

Orwell, G. (1945): *Animal Farm*, first published 17 August 1945, London, England: Secker and Warburg.

Orwell, G. (1949): *1984 — A Novel*, London: Secker & Warburg.

OWHA Clean Energy Investment (2018): "FACT SHEET: Obama Administration Announces Initiative to Scale Up Investment in Clean Energy Innovation," https://obamawhitehouse.archives.gov/the-press-office/2015/02/10/fact-sheet-obama-administration-announces-initiative-scale-investment-cl, released February 10, 2015; accessed April 25, 2018.

OWHA Clean Energy Savings (2018): "FACT SHEET: Obama Administration Announces Clean Energy Savings for All Americans Initiative," https://obamawhitehouse.archives.gov/the-press-office/2016/07/19/fact-sheet-obama-administration-announces-clean-energy-savings-all, released July 19, 2016; accessed April 25, 2018.

OWHA Climate (2018): "President Obama on Climate and Energy," https://obamawhitehouse.archives.gov/sites/obamawhitehouse.archives.gov/files/achievements/theRecord_climate_0.pdf, accessed April 25, 2018.

OWHA Recovery Act (2018): "The Recovery Act Made The Largest Single Investment In Clean Energy In History, Driving The Deployment Of Clean Energy, Promoting Energy Efficiency, And Supporting Manufacturing," https://obamawhitehouse.archives.gov/the-press-office/2016/02/25/fact-sheet-recovery-act-made-largest-single-investment-clean-energy, released February 25, 2016; accessed April 25, 2018.

Parshall, J. (2017): "Outlook Points to Peak Transport Demand for Oil," Journal of Petroleum Technology, September issue, pages 43–44.

Pease, E.R. (1916): *The History of the Fabian Society*, New York: E.P. Dutton & Company.

Petit, J.R., *et al*., 2001, Vostok Ice Core Data for 420,000 Years, IGBP PAGES/World Data Center for Paleoclimatology Data Contribution Series #2001-076. NOAA/NGDC Paleoclimatology Program, Boulder CO, USA; see also Vostok data at NOAA, Petit, J.R., *et al*., 2001, http://www.ncdc.noaa.gov/paleo/icecore/antarctica/vostok/vostok_data.html, accessed November 13, 2014.

Quigley, C. (1961): *The Evolution of Civilizations: An Introduction to Historical Analysis*, First Edition, New York: MacMillan.

Quigley, C. (1966): *Tragedy and Hope — A History of the World in Our Time*, New York: MacMillan.

Quigley, C. (1981): *The Anglo-American Establishment*, New York: Books in Focus.

RF — Work (2018): "Rockefeller Foundation: Our Work, "https://www.rockefellerfoundation.org/our-work/, accessed March 31, 2018.

Rizzi, B. (1939): *The Bureaucratization of the World* (1939 French edition translated by A. Buick from French to English), https://www.marxists.org/archive/rizzi/bureaucratisation/index.htm accessed October 7, 2017.

Roberts, S. (2015): "Maurice Strong, Environmental Champion, Dies at 86," published December 1, 2015 online at https://www.nytimes.com/2015/12/02/world/americas/maurice-strong-environmental-champion-dies-at-86.html, accessed March 21, 2018.

Rockefeller, D. (1991): "Foreward" in *Beyond Interdependence* by MacNeill, et al., 1991.

Rockefeller, D. (2002): *Memoirs*, New York: Random House.

Rockefeller, J.D., (1909): *John D. Rockefeller: The Autobiography of an Oil Titan and Philanthropist*.

Rockefeller, J.D., III (1973): *The Second American Revolution*, New York: HarperCollins.

Rosett, C. and G. Russell (2007): "At the United Nations, the Curious Career of Maurice Strong," http://www.foxnews.com/story/2007/02/08/at-united-nations-curious-career-maurice-strong.html, accessed March 25, 2018.

RSV Bible (1971): Genesis.

Sagan, C. (1980): Cosmos, TV mini-series documentary based on Sagan's book *Cosmos*, Norwalk, Connecticut: The Easton Press Collector's Edition, 2005.

Shafer, P.W. and J.H. Snow (1962): *The Turning of the Tides*, New Canaan, Connecticut: Long House Publishing.

Simkin, J. (2017): "James Keir Hardie," Spartacus Educational article, http://spartacus-educational.com/PRhardie.htm, accessed March 11, 2018.

Skousen, W.C. (1970): *The Naked Capitalist*, Salt Lake City, Utah: self-published.

Smil, V. (2003): *Energy at the Crossroads*, Cambridge, Massachusetts: The MIT Press.

Smil, V. (2015): *Power Density*, Cambridge, Massachusetts: The MIT Press.

Smith, H. (1991): *The World's Religions: Our Great Wisdom Traditions*, New York: HarperCollins; published by Easton Press with the permission of HarperCollins in 2008.

Solar Spectrum (2018): Figure annotated by the author and obtained from website https://commons.wikimedia.org/wiki/File:Solar_Spectrum.png, accessed 1-27-2018.

Strong, M. (1998): "A People's Earth Charter," Chairman of the Earth Council and Co-Chair of the Earth Charter Commission, March 5, 1998; http://www.earthcharterinaction.org/invent/images/uploads/Maurice%20Strong%20on%20A%20Peoples%20Earth%20Charter.pdf, accessed March 31, 2018.

Strong, M. (2000): *Where on Earth Are We Going?* Toronto: Alfred A. Knopf.

Sussman, B. (2010): *Climategate*, WND Books: Washington, D.C.

Sussman, B. (2012): *Eco-Tyranny*, WND Books: Washington, D.C.

Tarbell, I.M. (1904): *The History of the Standard Oil Company*, New York: McClure, Phillips and Company.

Torah (1992): Genesis, *Torah*, Norwalk, Connecticut: The Easton Press Collector's Edition.

Trotsky, L. (1939): "The USSR in War," first published in *The New International* (New York), Volume 5 Number 11, November 1939, pages 325–332; Leon Trotsky Internet Archive 2005; British edition published in Trotsky (1942).

Trotsky, L. (1942): *In Defense of Marxism*, first published in 1942 by Pioneer Publishers, online version transcribed by David Walters in 1998, https://www.marxists.org/archive/trotsky/idom/dm/dom.pdf accessed Oct. 7, 2017.

U Thant (1970): "Human environment and world order," International Journal of Environmental Studies, Volume 1:1-4, pages 13–17, DOI: 10.1080/00207237008709390.

UGS Ice Ages (2018): Utah Geological Survey, http://geology.utah.gov/surveynotes/gladasked/gladice_ages.htm; accessed 1-27-2018.

UNEP — Strong (2018): "Maurice F. Strong, UNEP Executive Director, 1972 to 1975," United Nations Environment Programme website, http://web.

unep.org/exhibit/UNEP-Executive-directors/maurice-f-strong, accessed March 24, 2018.

UNDP 2015 HDI (2018): United Nations Human Development Index using 2015 data, http://hdr.undp.org/en/countries accessed 1-25-2018.

UNDP 2016 HDR (2017): United Nations Development Program 2016 Human Development Report, http://hdr.undp.org/en/countries accessed 1-25-2018.

UNDP 2015 EIA (2018): United States Energy Information Administration, annual primary energy consumption data for 2015, http://www.eia.doe.gov/, accessed 1-25-2018.

UN DESA (2017): "World Population Prospects: The 2017 Revision," United Nations Department of Economic and Social Affairs, Population Division, DVD Edition, https://esa.un.org/unpd/wpp/Download/Standard/Population/ accessed January 27, 2018.

US AEO (2011): 2011 Annual Energy Outlook, United States Energy Information Administration, available at http://www.eia.doe.gov/.

US AEO (2017): 2017 Annual Energy Outlook, United States Energy Information Administration, available at http://www.eia.doe.gov/.

US DoE Sequestration (1999): "Carbon Sequestration: State of the Science," U.S. DoE Sequestration Report, Feb.

US EIA AEO (2017): "AEO2017 Levelized Costs," Annual Energy Outlook, United States Energy Information Administration, http://www.eia.doe.gov/, accessed 1-26-2017.

US EIA AER (2012): Annual Energy Review, United States Energy Information Administration, http://www.eia.doe.gov/, accessed 1-19-2012.

US EIA AER (2016): April Monthly Energy Review, United States Energy Information Administration, Tables 1.3, 2.1 to 2.6.

US EIA (2018): United States Energy Information Administration, http://www.eia.doe.gov/, accessed 1-26-2018.

US EPA Clean Power Plan (2018): "FACT SHEET: Clean Power Plan," United States Environmental Protection Agency, https://archive.epa.gov/epa/cleanpowerplan/fact-sheet-clean-power-plan.html#print; accessed April 25, 2018.

US EPA Sea Level (2014): United States Environmental Protection Agency, http://www.epa.gov/climatestudents/impacts/signs/sea-level.html; accessed November 27, 2014.

US IEO (2018): United States International Energy Outlook, http://www.eia.doe.gov/, accessed January 25, 2018.

US NOAA Arctic Ice Sheet (2018): United States National Oceanic and Atmospheric Administration Paleoclimatology Program, http://www.learner.org/courses/envsci/visual/visual.php?shortname=ice_sheet; accessed January 27, 2018.

US NOAA Keeling (2015): United States National Oceanic and Atmospheric Administration Earth System's Research Laboratory, image from NOAA's Global Monitoring Division website, http://www.esrl.noaa.gov/gmd/dv/iadv/; accessed January 31, 2015.

US NOAA NHC (2014): United States National Oceanic and Atmospheric Administration National Hurricane Center website, http://www.nhc.noaa.gov/surge/; accessed November 27, 2014.

US NOAA Solar Insolation (2018): Figure provided by U.S. National Oceanic and Atmospheric Administration, from website http://climate.nasa.gov/faq/ accessed January 27, 2018.

US NOAA Temperature (2018): "Climate Projections," United States National Oceanic and Atmospheric Administration Global Climate Dashboard, https://www.climate.gov/maps-data; accessed April 26, 2018.

US GCRP (2009): US Global Change Research Project, website http://www.epa.gov/climatechange/science/future.html; accessed November 2014].

Voosen, P. (2018): "Meet Vaclav Smil, the man who has quietly shaped how the world thinks about energy," interview published March 21, 2018 at http://www.sciencemag.org/news/2018/03/meet-vaclav-smil-man-who-has-quietly-shaped-how-world-thinks-about-energy?utm_campaign=news_weekly_2018-03-23&et_rid=344255017&et_cid=1926126, accessed March 26, 2018.

Vostok Temperature and GHG (2018): Accessed January 27, 2018 at http://www.iceandclimate.nbi.ku.dk/research/past_atmos/composition_greenhouse/.

WCED (World Commission on Environment and Development), Brundtland, G., Chairwoman (1987): *Our Common Future*, Oxford University Press.

Wells, H.G. (1909): *New Worlds for Old*, New York: MacMillan.

Whitman, A. (1974): "Walter Lippmann, Political Analyst, Dead at 85," accessed March 13, 2018 at https://www.nytimes.com/1974/12/15/archives/walter-lippmann-political-analyst-dead-at-85-walter-lippmann.html.

Wigley, T.M.L., R. Richels, and J.A. Edmonds (1996): "Economic and environmental choices in the stabilization of atmospheric CO2 concentrations," *Nature* (18 January), pages 240–243.

Wilson, W. (1918): President Woodrow Wilson's Fourteen Points, http://avalon.law.yale.edu/20th_century/wilson14.asp, accessed March 14, 2018.

Winkelmann, R., A. Levermann, A. Ridgwell, and K. Caldeira (2015): "Combustion of available fossil fuel resources sufficient to eliminate the Antarctic Ice Sheet," *Science Advances*, Volume 1, Number 8, e1500589, DOI: 10.1126/sciadv.1500589, September 11 2015.

Woolf, L.S. (1916): *International Government*, New York: Brentano's, website https://ia801406.us.archive.org/8/items/internationalgo00commgoog/internationalgo00commgoog.pdf accessed March 11, 2018.

World Bank (2018): "Our Mission," http://www.worldbank.org/ accessed April 1, 2018.

Yergin, D. (1992): *The Prize*, New York: Touchstone Book.

Yergin, D. (2011): *The Quest*, New York: Penguin Press.

Yergin, D. (2013): "EXECUTIVE PERSPECTIVE: Daniel Yergin on the puzzle of energy transitions," article published March 2013 for World Economic Forum, Energy Transitions: Past and Future — Energy Vision 2013; https://blogs.thomsonreuters.com/sustainability/2013/03/13/executive-perspective-daniel-yergin-on-the-puzzle-of-energy-transitions/, accessed April 14, 2018.

Yergin, D. (2015): "Energy Transitions: Present and Future," October 2015 Strategic Report, IHS Energy, http://cdn.ihs.com/energy/IHS-EnergyTransitions-Yergin.pdf, accessed April 15, 2018.

Index

A
absorption • 33, 34, 48
Africa • 27, 49, 93, 110–113, 136, 137
Agenda 21 • 141, 143, 144, 175, 177
albedo • 43
anarchy • 176
Antarctic • 37, 48, 58
anthropogenic climate change (ACC) • 24, 40, 46, 60, 70, 84, 145, 151, 166, 171, 172, 174, 177
Arctic • 47, 58
Asia • 89, 93, 117
Australia • 27, 89

B
Bilderberg group • 161
Brazil • 61, 141
Brexit • 177
Brooks, Arthur • 126
Browner, Carol • 167
Brundtland, Gro • 10, 141, 142, 145
Burnham, James • 106, 107, 115–117

C
Canada • 27, 113, 135–138, 141
capitalism • 84, 100, 115–117, 175, 176
Carnegie, Andrew • 155, 161
carbon capture and storage (CCS) • 39, 94
carbon capture, utilization and storage (CCUS) • 39
carbon tax • 80
Carson, Rachel • 133
catastrophic climate change • 65, 77
central planning • 84, 159, 175–177
Chaffetz, Jason • 107, 120
chaos • 88, 89
China • 50–52, 61, 74, 89, 90, 105, 111, 137, 146, 170
clash of civilizations • 87
climate change debate • 52, 58
climate model • 56–58, 78
Club of Rome • 162
Cold War • 16, 87, 88, 91, 133, 156
combustion • 14, 24, 28–30, 47, 48, 68, 71, 94, 95, 179
communism • 85, 88, 95, 137
conservation • 10, 92
conservative • 66, 159
conspiracy • 145
COP21 • 58–60, 77, 79, 170, 172, 175, 177

Council on Foreign Relations • 113, 114, 125, 161
Croly, Herbert • 129–131

D
Davidson, Thomas • 120, 121, 125
decarbonization • 64, 65, 69, 71, 178, 179
Declaration of Independence • 99
Deep State • 107, 118–120
Deepwater Horizon • 170
democracy • 85, 88, 162
Democrat • 166, 172
dialectical materialism • 96, 97, 102
drilling • 30, 37, 94

E
Earth Charter • 144, 145, 175, 177
eccentricity • 42, 43
ecology • 162
economics • 93, 127, 128, 158, 159, 175
economies • 11, 86, 92, 93
efficiency • 10, 13, 67, 72, 80, 152, 168–171
Egypt • 111
Ehrlich, Paul • 54, 55
Eisenhower, Dwight • 119
electric grid • 169
electric power • 5, 9
electricity consumption • 13
electricity generation • 21
electricity transmission • 65, 68
emission • 29, 38, 48, 52, 53, 64, 178
energy carrier • 179
energy conservation • 28, 92

energy consumption • 3–5, 11–14, 17, 27, 28, 50, 66, 72, 75, 76, 80, 178
energy industry • 58, 67
energy interdependence • 93
energy mix • 2, 3, 6–8, 10, 11, 65, 69, 73–76, 178, 179
energy production • 11, 28, 29, 64
energy storage • 68, 73
energy transition • 3, 68, 69, 71–75, 95, 106, 175, 178, 179
Engels, Friedrich • 96, 97, 100, 104
environmental impact • 3, 7, 14, 68, 75, 94, 100
environmentalism • 84, 95, 102, 105, 135, 139
Europe • 89, 101, 117, 124, 126, 132, 136, 137, 159, 160, 162
European Union (EU) • 49, 65, 68, 84, 90, 177, 179
evolutionary • 98
exploration • 45, 93

F
Fabian Society • 104, 105, 118, 120–126, 128, 129, 140, 158
fascism • 85, 159
Federal Energy Resources Board (FERB) • 67, 70, 80
Federal Reserve • 67, 70, 149, 150
feudalism • 162, 177
feudalist • 150
forecast • 26, 27, 38, 77, 80, 162, 175
fossil energy • 20
France • 49, 61, 80, 90, 105, 127, 131

frequency • 32, 34
Friedman, Thomas • 79, 80

G
Gaussian curve • 18
geologic sequestration • 48, 49
geopolitics • 87, 93, 95
Germany • 74, 90, 92, 101, 112, 127, 131, 132, 159
glacial • 44
glaciation • 47
global greening • 26
global warming • 16, 36, 58, 59, 145, 166
globalism • 84, 120, 124, 126, 135
globalization • 147, 151
Goldberg, Jonah • 106
Goldilocks Policy • 3, 68, 69, 75, 80, 81, 83, 84, 95, 106, 176–178
Gore, Al • 68, 166, 172
Grand Energy Bargain • 79, 80
Great Britain • 101, 105, 111
greenhouse effect • 24, 31–34, 42, 49
greenhouse gas • 14, 16, 21, 24, 26, 29, 33, 34, 36–38, 42, 48–50, 60, 61, 64, 69, 71, 86, 95, 168–170, 178, 179
Gross Domestic Product (GDP) • 13, 26, 85, 142
Gross National Income (GNI) • 11, 13
guano • 100, 101

H
Hardie, Keir • 122, 123
Hawaii • 14, 15, 35
Hayek, Friedrich • 159, 175, 177
Hegel, Georg • 95, 96, 99, 105, 106
hockey stick • 36, 46
Hofmeister, John • 28, 58, 67, 69, 70, 80
Holdren, John • 55, 167
Holland • 25
House, Edward M. • 67, 113, 122, 127, 130, 150, 161, 167, 168
Hubbert, M. King • 18, 20
Human Development Index (HDI) • 11–14, 27, 66
Huntington, Samuel • 87–90, 93
hydrate • 49
hydroelectric power • 6
hydrogen economy • 179
hydropower • 5, 8, 29

I
Ice age • 42–46
ice core • 37, 42, 46
Iceland • 13, 45
India • 113
Inquiry (The) • 126–128, 130
intangible cost • 65, 178
interglacial • 44, 45
internal combustion engine (ICE) • 22, 30, 76
international banker • 147, 150
internationalism • 158
International Thermonuclear Experimental Reactor (ITER) • 80, 81
IPCC • 35, 36, 45, 46, 55–57, 166
Iran • 91

J
Jackson, Lisa • 149, 167
Japan • 29, 111, 162

K
Keeling, Charles • 14
Keeling curve • 14, 15
Kerry, John • 60, 177
kinetic energy • 2, 34
Klare, Michael • 93
Koonin, Steven • 58
Korea • 146

L
Labour • 104, 122
Lankester, Edwin Ray • 102, 103
Laws of Matter •97, 99
League of Nations • 105, 114, 125–127, 130–132
Lenin, Vladimir • 103–106
levelized cost of energy (LCOE) • 21
liberal • 84, 128–130
Liberal-Progressive • 67–69, 172
Liebig, Justus • 100, 101, 103
Lippmann, Walter • 128–132, 150
Little Ice Age • 16, 45, 46
Locke, John • 99, 100
London School of Economics (LSE) • 128, 158, 159

M
MacDonald, Ramsay • 122
Malaysia • 89
Malthus, Thomas • 54
Managerial Revolution • 107, 115
managerialism • 117
Marx, Karl • 95–97, 99, 100–106, 121
mass media • 86
materialism • 96, 97, 107, 164
McCarthy, Regina • 167
Medieval Warm Period • 16, 45
Mexico • 61, 89
Middle East • 49, 90
Milankovitch cycle • 42, 43
Milankovitch, Milutin • 42
Milner, Alfred • 102, 112, 113, 125
monarchy • 85
monetary policy • 150

N
natural resources • 92, 100, 102–104, 134, 141, 142, 162, 168, 171
New Republic • 129–131
New Zealand • 89
NIPCC • 55–57
nonrenewable • 4, 5, 18, 22
North America • 55, 89, 161, 162, 177
nuclear fission • 2, 5, 6, 9, 22, 29, 49, 72, 79, 80, 179
nuclear fusion • 5, 80, 81, 179
nuclear winter • 16

O
Obama Administration • 55, 120, 167–172
Obama, Barack • 58, 67, 68, 119, 120, 166–172, 174
obliquity • 42
oil crisis • 93
oil price • 67, 93, 94

Oil and Gas Climate Initiative (OGCI) · 60, 61
oligarchy · 106
OPEC · 92
Orwell, George · 117, 118

P
Paris Climate Agreement · 59, 60, 77, 79, 170–172, 175
peak oil · 18, 22, 93, 94
Pease, Edward · 121–125
photosynthesis · 26
Plato · 96, 109
plutocracy · 106, 107
political obstacle · 84, 106, 120, 126
pollution · 35, 60, 86, 134, 162, 168, 170, 177
Population Bomb · 54, 55
population forecast · 26, 66
population growth · 26, 52–55, 66, 134, 144
potential energy · 2
precession · 43
privileged minority · 140, 151
progressive · 98, 105, 129, 130
property · 95, 96, 99–101, 109, 115, 117, 126, 152
public policy · 3, 69, 108, 179

Q
quality of life · 11, 13, 14, 17, 20, 27, 28, 55, 65, 66, 179
Quigley, Carroll · 106–108, 112, 147, 148, 150

R
radiation · 14, 32, 33
regulation · 55, 69, 70, 84, 98
republic · 85
Republican · 153, 166, 172
reserves · 148
revolutionary · 103, 104, 120, 140, 164
Rhodes, Cecil · 110–112, 161, 174
Rhodes Secret Society · 112, 161
Rio Conference · 141, 142, 144
Rizzi, Bruno · 106, 107, 115, 116
rock oil · 6, 8
Rockefeller, David · 136, 157, 158, 161–163, 165
Rockefeller, John D. · 153, 154, 157, 158, 163, 174
Rockefeller, John D., Jr. · 153, 154, 157, 158
Rockefeller, John D., III · 163
Rockefeller, Nelson · 154, 163, 165
Rockefeller Brothers Fund · 160, 161
Rockefeller Foundation · 153, 154, 156, 157, 163
Round Table · 113, 114, 161
rule of law · 146
ruling class · 84, 85, 99, 106–110, 115, 116, 175, 177
Ruskin, John · 108–110, 114
Russia · 89, 90, 103, 104, 131

S
Saudi Arabia · 61
sequestration · 48, 49
serfdom · 159, 177

Sherman Antitrust Act • 152
Skousen, Cleon • 147, 150
Smil, Vaclav • 13, 47, 48, 70–73, 162, 175
socialism • 84, 85, 95, 115, 117, 118, 120–122, 126, 128, 130, 132, 158, 159
socialist environmentalism • 95
solar energy • 5, 20, 28, 29, 33, 68, 71, 73
solar irradiance • 33
solar power • 30, 33, 49, 65, 86
solar radiation • 33, 34
South America • 89, 111
Spain • 61, 90
Stalin, Joseph • 105, 106, 115, 116
Standard Oil of New Jersey • 136
Starndard Oil Trust • 152
Stockholm Conference • 135, 138, 139, 141, 142
Strong, Maurice • 135, 137, 141, 144, 145, 147, 151, 157, 160, 192, 166, 174
sunlight • 26, 32–34, 71
supergrid • 92
Supranational Authority • 84, 124–126, 131, 132, 146
Sussman, Brian • 16, 45, 95, 99, 100
sustainable development • 10, 141, 143, 144
sustainable energy • 2, 3, 10, 24, 49, 65, 75, 77, 79, 81, 175, 177–179

T

Tansley, Arthur • 103
temperature change • 16, 29, 30, 35, 37–39, 44–46, 52, 53, 77–79, 175

Thant, U • 132–135, 138, 174
time scale • 43
tokamak • 80
Tragedy and Hope • 106–108
Trilateral Commission • 161, 162
trilateralism • 162
Trotsky, Leon • 115, 116
Trump Administration • 60, 172
Trump, Donald • 60, 67, 87, 120, 172, 177
Turkey • 89, 131

U

Ukraine • 49, 79
uncertainty • 3, 179
United Kingdom • 111
United Nations Environment Programme (UNEP) • 55, 56, 139, 141, 166

V

Viking • 45
Vita Nuova • 120, 125
volatile organic compound (VOC) • 14, 34
Vostok • 37–39, 41, 42

W

Wallas, Graham • 128, 129
whale oil • 6, 8, 30, 64
wind energy • 20, 75
wind farm • 29, 49
wind turbine • 21
Wilson, Woodrow • 105, 114, 122, 125–132, 149, 150
Wilson's 14 Points • 129–131

Woolf, Leonard • 124, 125
World Bank • 174
World War I • 6, 52, 102, 103, 105, 107, 113, 114, 124–128, 150
World War II • 6, 52, 88, 105, 107, 125, 132, 133, 156, 160, 161

Y

Yergin, Daniel • 56, 57, 73, 74, 151–153, 175
Yin-Yang • 97, 98